# The Quantum Internet

T0171894

Gösta Fürnkranz

# The Quantum Internet

## Ultrafast and Safe from Hackers

With a Foreword by Rupert Ursin

 Springer

Gösta Fürnkranz
Hinterbrühl, Austria

*Translated by*
Andrea Aglibut
Vienna, Austria

ISBN 978-3-030-42663-7          ISBN 978-3-030-42664-4    (eBook)
https://doi.org/10.1007/978-3-030-42664-4

This Springer imprint is published by the registered company Springer Nature Switzerland AG
The registered company address is: Gewerbestrasse 11, 6330 Cham, Switzerland

# Foreword

The vision of a quantum internet—everyone even remotely interested in technology will come across it time and again—is described by Gösta Fürnkranz as "ultrafast and safe from hackers". With his far-reaching, comprehensive and at the same time entertaining presentation of the status quo, the author fulfils an important task for the general public. Until now, we had to make do with journal articles and simplified excerpts from science publications to get an idea of these exciting new developments. Sure, physicists, like myself, focus primarily on specialized communication with our colleagues in the field. We are less concerned with comprehensibility, because the precision that comes with the technical jargon—which may come across like gobbledygook to non-scientists—brings us further in our research and development work. And yet I believe that every person today has the right—I would almost be inclined to speak of a duty if I did not find myself so negligent of this in so many other areas of

knowledge—to be up to date of important and momentous findings and developments. In the age of Facebook
and Twitter, the times when such knowledge was reserved
for a handful of scholars inhabiting monastic ivory towers
are finally over, and that is a good thing. Unfortunately,
quality and accuracy of the information distributed via
the so-called new media all too often fall by the wayside.
I notice this again and again with a queasy feeling and a
touch of a guilty conscience. I always resolve to pay more
attention to comprehensibility and ease of communication, but all too often I do not get the chance to put
those good intentions into practice in my everyday life as a
researcher between experiments and writing articles, giving
lectures and attending meetings in commissions, commit-
tees and whatever else science management demands.

All the more I am happy about the comprehensive,
in-depth and at the same time entertaining and easy to
read presentation by Gösta Fürnkranz. The author quite
deliberately renounces the frequently quite arrogant love
of detail to which we scientists often devote ourselves.
In the—on occasion quite lively—discussions with the
author, I was often convinced, if not corrected, that what
is important is understanding the big picture and not to
communicate every tiny technical detail. In the end, you
do not need a dictionary app for this book; you will be
fine even reading it in the bathtub. After your perusal of
this book, you, dear reader, will know what the quantum
internet is all about. And that is not all. You will know
much of what there is to know about the history and present of quantum physics, the current state of communication technologies and the possibilities that result from it
for the future. The author also introduces his readers to
the beauty of the scientific method without neglecting its
limitations.

Fürnkranz skillfully combines the complex mosaic of foundational principles of physics, the technological background, historical examples, cutting edge experiments and possible scenarios into an intuitively comprehensible overall picture of quantum communication technologies. With a light touch, he brings to life historical figures of quantum physics as well as today's pioneers of quantum information. A special treat by the full-time teacher is the workshop part of the book. Here, fans of precision get their money's worth. If you read this, you will get a comprehensive overview of the underlying quantum physics—without having to attend the full curriculum of a physics student.

Together with the author, I would like to recommend one of the basic principles of the scientific method to all readers: falsifiability. What we are trying to do in science is to make sense of the world, to derive all possible hypotheses from what has been learned and found so far and thus build up an ever new understanding of the world. And then, we take a step back and take a look. Where can we cut with the sharp knife of the science experiment? Which always carries the risk—or the opportunity, depending on the way you look at it—of falsification, while never being able to bring about true verification. And if a hypothesis has been falsified in an experiment (by facts), then it must be abandoned, then it cannot be saved, even if we have put a lot of work into it and have formed an attachment to it. With what remains after the experiment, we may continue our work. These are our hypotheses, and in the end, they are always only working hypotheses. And these working hypotheses are what is considered the state of the art. In today's physics, where we try to combine quantum theory with other disciplines in physics and at the same time apply it to new technologies, this process is particularly rapid and explosive. Fürnkranz describes in detail and very comprehensibly the current state of research and development. However, this also means that details or even entire areas can be superseded by newer developments in just a few years or even months. The book you are currently reading vividly reflects this vitality of current physics.

I was very pleased to be invited to write this foreword and to have the opportunity to read this interesting book before all other readers. I confess that I was pleasantly surprised by the quality and topicality of this contribution. I had not expected this from someone who in his everyday life teaches at a secondary technical school, someone who in no way actively conducts scientific research. I would like to thank Gösta Fürnkranz for this very important contribution to communicating science to a very broad readership. It is (and I say this with particular emphasis, precisely because this book was written by a teacher) one of the best representations of the subject area in recent years, including all specialist publications and survey articles by leading scientists.

It only remains for me now to wish you, dear readers, much fun and a whole lot of eye-openers during your reading.

Dr. Rupert Ursin
Vice Director
Institute of Quantum Optics
and Quantum Information Vienna
(IQOQI Vienna)
Austrian Academy of Sciences
Vienna, Austria

# Introductory Remarks

For quite some time now, digitization has been one of the determining factors of the world we live in. Its future development offers a variety of diverse opportunities and possibilities. Now is the time for action. Among its most important aspects is secure digital communication. To protect digital safety in the long run, innovative technologies play a decisive role. Instances of unauthorized access and manipulation of communication networks are increasing rapidly. More and more damage is done to individuals, society and economy. The pervasive use of the internet creates an enormous threat potential of criminal and terrorist activities. Issues of data security and integrity are rapidly gaining significance, especially in view of the steady growth of online businesses, systemic networking and the progress of the internet of things. Many of today's keywords, for example, Industry 4.0, autonomous driving, wearables or smart city, are harbingers of a fully networked digital society where the quantity of data will explode exponentially. In the long term, this development can only be countered by the development of completely safe technologies. Almost without exception, security technologies up to now have been based on merely making access to data more difficult. There is, however, an alternative solution which renders unauthorized access to data and information completely impossible—for reasons inherent in the physical laws of nature: quantum communication.

At the heart of this development is the technology of quantum information, which opens a radically new approach to information theory. In contemporary IT, data processing and transfer are performed exclusively on the basis of bits and bytes, i.e. sequences of binary numbers which, by definition, only contain the digits 0 and 1. Quantum information, on the other hand, defines the quantum bit (qubit) as its elementary unit. The qubit represents a kind of simultaneous superposition of 0 and 1. This has two decisive advantages over the classical concept of information. On the one hand, quantum bits can store and transmit substantially larger amounts of information than conventional bits. On the other hand, qubits contain an inherent safety feature that makes it impossible to intercept or hack quantum information. This entirely new functionality has no counterpart in classical IT. Classical information can be copied and duplicated at will and is therefore inevitably vulnerable to unauthorized distribution. For fundamental reasons inherent in the laws of nature, quantum bits are immune against such attacks.

In recent years, fundamental research has revealed a number of essential underlying principles that confirm the great potential of this technology. Recent groundbreaking experiments have paved the way for the development of quantum communication technologies. One example is the successful realization of quantum channels over distances of up to 1203 km. First, quantum cryptography devices are being introduced to the market by new companies. Considerable efforts worldwide are focused on developing the technological foundations for global distribution. These procedures aim to facilitate quantum communication over long distances and to multiuser access. Ultimately, this requires a special quantum network, early prototypes of which were first implemented in Europe, Asia and the USA. In China and Europe, significant funding instruments were created specifically to support active research by institutions and companies in this field. Experts predict a bright future for this development. European research (where this technology emerged first) is particularly interested in developing tap-proof quantum cryptography to market maturity.

Another driving force behind the development of future quantum networks is the current state of computer technology, which for fundamental reasons will reach its limits in the foreseeable future. An additional challenge is that already today, a number of IT problems exist that even supercomputers are not able to solve on reasonable timescales, or at all. The search for novel concepts was started in the computing world a long time ago. The most revolutionary approach to date is the concept of the quantum computer. The most ambitious goal of quantum informatics is the development of a technically feasible computing machine based on the laws of quantum physics. The considerable commercial interest of industry, including first and foremost global players such as Google, IBM or Microsoft, suggests that this goal is a realistic one. Recent studies by Morgan Stanley also attach great importance to the quantum computer, and even critics like the computer scientist Scott Aaronson agree. Although currently still in an early stage, quantum computers are possibly the only hope to significantly improve the computer performance of classical computers. If the quantum computer should ever cross the threshold into technical feasibility, it offers the prospect of an even more fascinating vision. The combination of inherent data security and the potential networking of quantum computers might one day result in the emergence of a quantum internet. And that would lead to a radical paradigm shift in terms of security and processing speed. Due to the theoretical capacity of quantum computers to answer certain problems that today cannot be solved even by supercomputers, such a quantum hypernet might provide undreamt-of possibilities for the future world of information.

The author would like to add a few words on the structure and interpretation of the current book. I have kept my wording deliberately optimistic. We have every reason to be hopeful. Scientific research has achieved numerous relevant breakthroughs, and it is certainly worthwhile to present the perspectives and potentials of this young field of quantum technologies to a wider audience. Especially in view of the fact that the science involved is so fundamentally important, with amazing consequences for the way in which we see the world. Of course, quantum physics remains controversial and subject to heated debates even among its most knowledgeable experts. And in popular science, the balancing act between scholarly precision and simplification for the sake of easier understanding is a particularly challenging one. I would, therefore, ask my readers to read the book in the sequence it is presented, as it relies on an almost universal didactic structure. Numerous science terms are mentioned in the early parts of the book and then revisited again and again to be explained in more detail and depth. Already in Chap. 1, experimental arrangements are introduced which form an important basis for experiments and concepts that are discussed in Chaps. 2 and 3. In some respects, the author deliberately departs from the traditional models of explanation. This is done in order to introduce the concept of information (in the sense of a deeper physical entity), which has been receiving increased attention recently. To some extent, my text supports interesting positions represented by leading experts in the field (e.g. Anton Zeilinger). In this sense, the author would like to provide comprehensive information about the current state of research and the latest developments on the path to a future quantum internet. Let me take you on a fascinating journey into the future and present you with a sneak preview of a new technological era which one day might turn out to be our reality.

# Contents

# 1

# The Quantum Digital Future

## 1.1 Digital Visions

When the industrial revolution began about 200 years ago, it brought about global change. It came hand in hand with a profound transformation of economic and social conditions, which greatly accelerated the development of productivity, technology and science. On the downside, it led to a number of social problems associated with workers' discontent that made new regulations and social reforms necessary. Now, in the twenty-first century, humanity is facing similarly epochal changes. Back then, the steam engine replaced muscle power. Now, in the digital age, the microchip is about to make mental labor redundant. What began in the 1940s with the development of the computer, later enabled the first lunar landing and led to the rise of pocket calculators and home PCs, now culminates in the development of the internet and its mobile devices. This also marks the beginning of

© Springer Nature Switzerland AG 2020
G. Fürnkranz, *The Quantum Internet*,
https://doi.org/10.1007/978-3-030-42664-4_1

the information age, the future direction of which stipulates the total networking of everyone with everything. Currently, the internet connects billions of people. It is expected to soon comprise some 40 billion networked devices. With tremendous dynamical power, digitization opens a new chapter in human development. Digital infrastructures, products and services are changing society and economy. This transition to a new modernity is commonly labeled the "digital revolution"—a process that is far from complete. This is particularly true for the internet of things, where futurologists see great potential in portable electronics, technical assistance systems, robotics and artificial intelligence. This is associated with modern, systemically networked production processes to increase efficiency and innovation. Further major changes are emerging in the area of mobility, where the focus is on digital networking of public transport and autonomous driving.

As history teaches us, technological development can be a powerful motor for social change—both in positive and in negative ways. New technologies have always confronted humanity with challenges. They have expanded our scope for action for better or for worse, they have made our lives easier or enabled destruction. This has been the result of progress through technological change from the Neolithic Revolution to the Industrial Revolution. One example is the invention of the printing press, which radically changed not only science, but the way we see and experience the world. Digital change poses new challenges and threats. The dangers of complete surveillance and restriction of personal freedom have to be considered, but protection against cybercrime and ethical issues relating to artificial intelligence need to be taken into account. The fact that new technologies replace masses of workers has accompanied every technological change to date. On the other hand, new fields of activity are emerging continuously. Many companies will need to embrace change in order not to get swallowed up by digital disruption (replacement of existing products and structures with new technologies and systems). This is why political guidelines include statutory regulations that set modern framework conditions and safeguard social security, to enable employees to realize the positive potential inherent in the technology.

The continuous progress in microelectronics and communication technology brings to life the vision of an all-encompassing network of countless sensors and computers, embedded in one's personal environment. Tiny processors, memory devices and low-cost sensors can be implemented in numerous everyday objects and appliances. Not only have microprocessors become smaller, more powerful and less expensive over recent decades. Also, wireless sensors make it possible to monitor and diagnose systems quickly and inexpensively from a distance. They can be installed and adapted in large numbers without the need for expensive cable connections, and integrated unnoticeably into objects that previously had not been network compatible. In conjunction with location recognition capabilities, such wireless devices achieve unprecedented quality. The pervasive smartphone culture, but also radio frequency identification tags or chips in ID cards and credit cards are harbingers of a new era of "ubiquitous computing". As early as 1990, the computer engineer and communication scientist Mark Weiser predicted: "In the twenty-first century the technology revolution will move into the everyday, the small and the invisible" (https://de.wikipedia.org/wiki/Ubiquitous_computing). As a reaction, the term "ambient intelligence" was coined in Europe, which focuses on the digital communication between everyday objects to improve and simplify our lives. Research in this area is aimed at networking processors, sensors and radio modules in such a way that they react adaptively to the needs of their users.

At the same time any visible technology is to melt into the background, functioning in almost imperceptible ways. For example, the presence of different persons is detected by systems in their environment, enabling the technology to react individually and unobtrusively. Everyday objects are to transform from passive things into active appliances and flexibly adapt their service for different users. Innovative interfaces such as speech or gesture recognition are of enormous benefit for this endeavor. In the long run, ambient intelligence is to encompass all areas of life. Any smart home of the future furnished with such technology will increase comfort and protection, but also support the optimization of energy management. In the office, productivity will be increased, and effectiveness will be improved with the help of smart assistance systems. In the area of intelligent transport, ambient intelligence will make traffic safer and help to conserve resources. Also, sensor networks are able to perform comprehensive monitoring tasks. Of course, it is important to keep a sense of proportion, so that the average citizen won't end up exposed to total monitoring.

The 5th generation of mobile radio standards (5G expansion) is essential for the future use of the internet. Data rates of up to 10 Gbit/s and low latency times make high densities of mobile devices possible. This opens up a multitude of new business models and applications. This "hyper-networked 5G era" is expected to encompass more than 40 billion networked terminals by the 2020s. This creates a solid foundation for the internet of things, which supports ambient intelligence. Devices communicate with each other and provide additional information on the internet. Instructed with the needs of the users, these appliances provide automatic support. Industry benefits from better machine maintenance, for example by means of automatic communication of status information. Another category concerns wearable devices which, for example, record vital signs (such as heartbeat or blood pressure) and transmit relevant data to medical centers, supporting remote monitoring of a patient's medical condition. Augmented and virtual reality systems have the potential to convey unexpected impulses, for example by displaying additional visual information or objects in real time via special glasses, essentially creating an interactive virtual environment. Any number of possible applications become imaginable, from tourism and education to craftwork or the construction sector. For example, a building project may be viewed in a virtual space before construction has even begun. Or, work instructions might be imported directly into an object.

The internet of things also forms the basis for autonomous driving. This particular challenge is increasingly shifting into the focus of the automotive industry. It facilitates new ideas for integrating and optimizing public and private transport. This results both in greater comfort and reduced environmental impact. New concepts support the prevention of accidents, the alleviation of parking problems, the mitigation of traffic congestions and not least the reduction of active vehicle owners. Mainly realized as an assistance system today, this technology will at some point develop into fully autonomous driving. 5G expansion plays a major role, as large amounts of vehicle data have to be transmitted and processed in split seconds, something that poses enormous challenges for mobile network operators. Very precise cartographical material that is constantly updated becomes a must, in addition to the simultaneous recording of the vehicle's position. Further data requirements include route, road conditions, current traffic situation, weather conditions, driving maneuvers of other cars and much more. This also creates an immediate competence problem: Who actually owns these data and how may they be used? Another aspect of course concerns hacker and software security, which obviously has to be on a very advanced standard. Likewise, completely new legal questions arise, such as the legitimacy of claims in the event of an insured event occurring. Would the "driver" be completely exempt from sanctions and liability in such cases? Who would be responsible instead?

The catchword "Industry 4.0" refers to the industrial use of modern information and production technologies that are to be connected in this way. Intelligent and digitally networked systems serve as the basis for this. This will to a considerable degree enable the automated management of production. People, machines, facilities and logistics, and even the products themselves, will cooperate and communicate directly with each other. This integrated network is to support the optimization not only of discrete production steps, but of an entire value chain, whereby all phases in the life cycle of the product are covered, including recycling. Industry 4.0 is often understood as a project for the future based on the networking of machines with sensors and functional transparency, i.e. expansion through sensor data, technical assistance and decentralized decisions. However, to arrive at this level of technology, numerous challenges have to be addressed. The main objective is to merge IT and production technology. At the core of this is a so-called cyber-physical system, i.e. a network of software components with mechatronic parts that communicate with each other via an infrastructure (e.g. the internet). On the basis of standards and norms, innovative products and services are to be expected. In this context, data as a "new raw material" is of particular importance, with data security and ownership naturally playing a key role.

The recent progress in computer technology and the explosive increase in the amount of information generated by networking are creating new perspectives for further progress in artificial intelligence (AI). For a long time, the topic of AI has slowly entered into the focus of companies and the public. Possibilities for application are diverse and include manufacturing, maintenance, logistics, sales, marketing and controlling, but also search algorithms and much more. Even today, computers are capable of processing unstructured information (e.g. speech or photos) in addition to structured information. This makes it possible to generate and process additional data that had previously been inaccessible. Besides, machine learning is increasingly gaining importance. Computers learn from each individual case. This reduces the probability of errors even further and optimizes action sequences. Besides its use in industrial applications, a modern robot is able to diagnose a tumor in just a few minutes. One day, neuro-prostheses will become possible, i.e. neuronal parts will replace motor, sensory or cognitive abilities that have been impaired by injury or illness. Beyond classical computer science, innovative concepts such as quantum computers could lead to completely new possibilities and perspectives in machine learning. Some experts believe that quantum processors will revolutionize machine learning. Companies such as Google, IBM and Microsoft are already investing in the vision of merging AI with quantum computing. Here, too, ethical questions are becoming more and more important. In all disruptive technologies, such questions must be addressed. When introducing AI into companies, the concern that jobs will be lost due to technical progress troubles employees already today. Management can alleviate such fears by convincing their staff that in most cases AI can only unfold its full potential through interaction with people.

With intelligent power grids and smart grids, future demands for economic-ecological optimization will be met. These enable direct communication between consumers and network operators, which balances supply and demand in the distribution network and promotes a sustainable transition to renewable energies. One example is the generation of electricity from wind power or photovoltaics, which is subject to natural fluctuations. The intelligent power grid reacts adaptively by coordinating the interaction of consumer, generator and storage through digital communication in such a way that the best possible efficiency is guaranteed. This fulfills an important prerequisite for the vision of future smart cities, which focuses on employing digital technologies for the efficient use of sustainable sources.

As a further means for the long-term conservation of energy and resources, 3D printing technology is predicted to have a great future. This field, which is also backed by corporate groups, is becoming more and more interesting for complex applications. It might even replace standard manufacturing processes at some point in the future. Already today, houses in Asia exist that were 3D printed rather than constructed in the traditional way. Accordingly, production systems can be decentralized, so that production and consumption take place at the same location. It is still undecided precisely what effect this would have on sales, distribution and transport if this technology penetrates the market. On the other hand, increases in economic efficiency are considered to be proven as compared to many competing manufacturing processes. Besides, effectiveness increases as the component geometry develops into expanding complexity. Keywords such as "bioprinter" or "digital food" promise that one day we will see significant innovations in health care and food production. It is conceivable that online shops will use this technology so that customers no longer purchase physical goods but rather download digital design plans which they then feed into their private 3D printers. In any case, the data material involved is extremely complex and needs to be protected accordingly.

## Future Products: Data Protection and Processor Performance

In view of the pervasive trend towards networking and the visions outlined above (which didn't arise out of the author's imagination, but have already been widely discussed), it is obvious that digital security in future IT has to be considered much more thoroughly than has been the case. This is true not only for our internet of communication, but also for the internet of things, which—as numerous experts predict—will become more and more pervasive in coming years. From a global perspective, even today millions of hackers and eavesdroppers are descending onto the internet with every second, causing enormous economic damage. In an ever more networked world, this problem will undoubtedly escalate. We don't even want to imagine what this might imply for fully autonomous driving operations in the future. A well-targeted cyber-attack on a managing system that controls tens of thousands of cars may lead to downright catastrophic results. It goes without saying that critical infrastructures, especially in connection with modern industry systems, need special protection. A fundamental problem is that the amount of generated data (which grows exponentially each year) will reach exorbitant proportions in the future. Not only is the risk of unauthorized and criminal attacks increasing rapidly, the sheer amount of personal data is dizzying as well. The volume of information already generated on the internet today (about 200 exabytes per month by 2020) is far too substantial and complex to be processed by conventional means. For this reason, large amounts of data are often collected centrally and then interconnected (big data). This is useful for many purposes, including business, finance and medicine. On the other hand, with the accumulation of ever-growing amounts of personal data, the protection of privacy and data sovereignty becomes an ever-increasing challenge. In such a highly interconnected world, "genuine privacy" has to be regarded as one of society's most important demands. Otherwise, the threat of the total surveillance state (which is already emerging in some places) looms at the horizon for all of us.

As current examples show, transactions with personal data are a flourishing business, which can lead to legal requirements being conveniently "forgotten". This calls for long-term protection systems, which should not, however, consist exclusively of regulations, but also include technical safety measures. Data is often portrayed as the gold of the future, and its analysis and distribution generates huge amounts of money for large business. It is plausible that at some point, opposing business models will emerge. Comprehensive cyber protection and privacy therefore needs to be taken seriously as a key business factor in the future. Of vital importance are also issues of central data storage, digital archives and database systems where a great amount of material is stored already today. What is considered safe now has to meet emerging security needs also in 20, 50 or 100 years. Banks and large companies have predominantly taken the view that although today's security technology is regarded to be sufficient, on a certain day X in the future, this might no longer be the case. And with that comes a certain uneasiness. It has to be stated explicitly that digital security technology is based exclusively on the assumption that the attacker's computer performance is not sufficient to crack the code, or the existing firewall. However, there is no direct scientific evidence to support this assumption. One weakness of today's public key algorithms (such as RSA or elliptic curves), which are used for digital signatures or key exchange, is that they are based on the complexity of mathematical problems. Breakthroughs in research and the constant increase in computing power can lead to these algorithms being cracked. The basic question is therefore: How can we safeguard long-term and sustainable cyber protection with the ability to withstand future computer developments of potentially very high performance?

It is therefore in the interest of society as a whole (and not just governments and elites) that science develops new concepts for digital security. Quantum communication is the ideal prerequisite for this. This concerns specifically the innovative approach of inherent security, where a system's effectiveness is not affected by the attacker's computer performance but contains a fundamental mechanism, based on the laws of physics, that guarantees immunity. Based on information technology to date, such a physics-based procedure is not possible. Quantum communication, on the other hand, offers a way to automatically close a crucial potential security gap: the completely tap-proof data connection between two remote points. Such a high-security connection can be established either directly from point to point or distributed through trusted nodes. In combination with classical security technology methods, it can also provide unprecedented protection against hacker attacks and unauthorized access to databases. Actually, this technology, in its fundamental principles, is complete and very close to market maturity. For this last step, sufficiently large investments are necessary. In Asian test networks, this technology has been implemented on a very large scale. US initiatives aim to get a hacker-proof quantum network that people can use. Some expect such networks to achieve supra-regional distribution in the 2020s. The technology may play an important role in local structures as well as in backbone networks. As this also involves numerous commercial applications, it might ultimately culminate in a global high-security network which is constantly being further developed and would therefore be ideally suited to future security requirements. Already today, companies offer security solutions based on quantum key distribution (QKD) that improve the safety of traditional cryptography systems. Such systems combine distribution appliances with link encryptors connected by optical fibers. Typical applications include secure LAN extensions, enterprise environments or data center links. Connection bandwidths of up to 10 Gbit/s and ranges of up to 100 km facilitate their use in metropolitan quantum networks. It is conceivable that numerous consumers will apply quantum modules in their computers to achieve tap-proof communication in the near future.

A glimpse into the digital future seems to reveal a rapid increase in computing performance. This is not only a consequence of the above-mentioned exponential data growth and the increasing processor power this necessitates; it also concerns future logistics and optimization tasks. As can be shown, numerous problems exist which cannot be solved by classical computers at all, or at least not within reasonable time frames. A well-known example is the problem of the traveling salesman. Determining the shortest route connecting all required destinations seems to be a considerable challenge for any conventional computer. After all, it is necessary to identify the best path possible from quintillions of viable variants (and this is true already if our salesman wishes to visit no more than 20 cities). A global problem of the future which is related to this concerns for example the optimization of traffic flow for autonomous driving systems. To gather extremely large amounts of data using sensors is not a technical problem, the subsequent simultaneous calculation of best possible driving maneuvers for all vehicles is. Relying on conventional EDP methods, the time frame classical computers require to do this is far too long to be of any use for real-world applications. As first simulations done by quantum computers suggest, such and similar optimization tasks can be achieved in much shorter periods of time with this new concept. Besides, there are numerous further logistical challenges. First and foremost, however, there's a large number of problems in science that cannot be solved with the help of classical computers, or at least not within a reasonable amount of time. New computer concepts are also gaining importance in view of AI and machine learning, not least because the "silicon revolution" we know so well will very likely be exhausted within a few years. Out of all potential technologies, the quantum computer is the most promising concept. It is probably the only way to significantly improve computer performance or even take it to a new dimension. For example, quantum computers are able to solve the very complex combinatorial optimization problems AI applications are faced with much more efficiently. They are also able to perform pattern recognition tasks with noisy data much more rapidly, in this way providing new perspectives for machine learning. Already today, it is evident that any arbitrary digital quantum simulation of a complex problem can be performed with the help of quantum simulators. IT giants including Google, Microsoft or IBM, who already invested billions in these new technologies, demonstrate the great market potential. Volkswagen, for example, has entered into a cooperation agreement with Google to have calculations for batteries and autonomous vehicles created using quantum processors. This implies that surely, quantum technologies will also play an important future role from this perspective. While its value for science is enormous, the development of technologically serviceable quantum computers and the connected network technology is also in the focus of research. The QKD internet already represents a tangible goal with clear contours (governments and companies have already articulated their interest). A network of powerful quantum processors, however, still remains a pure vision of the future. It is not even clear which functionalities exactly might be integrated. Similarly, it is not yet possible to estimate to which extent its realization is physically/technologically possible at all. The considerations that have to be considered are manifold. For some researchers, all this still remains predominantly a grey area. Other experts are already indulging in fascinating speculations about an unspeakably powerful hypernet that will one day emerge out of these new quantum technologies, which will set entirely new standards for system coordination, processing speed, data rate and security. It is important to note explicitly that the speed advantage of a quantum internet is not the result of an extremely fast transmission of directly usable information. Rather, it owes its processing speed to the fact that quantum bits are able to store and transfer much larger amounts of information than conventional bits. Even though the quantum internet is not able to communicate faster than light, it will be able to coordinate and synchronize tasks at superluminal velocities. In this, it is truly unique (such and similar aspects will be discussed in detail later). After all, a quantum internet would not only benefit from the acceleration of its quantum processors. If configured accordingly, it would also be able to contribute to the development of scalable quantum computers (i.e. expanding their processing power to an arbitrary number of qubits). In view of the requirements of our future's internet, it is important to note the following. If the ideas envisioned for classical IT (which would not be realized immediately, but in a kind of gradual revolution) are to be meaningful in the long run, the future innovations in conventional IT will need to be complemented by "quantum digital" technologies. This is true not least because quantum computers theoretically present a danger to traditional security technology, making new safety methods necessary. Ironically, these measures are in turn largely based on quantum theory.

## 1.2    Revolutionary Quantum Physics

Quantum technologies were covered by European Forum Alpbach's Technology Symposium several times. For many decades, quantum technologies have had a revolutionary impact on humankind. Technical innovations including lasers, imaging techniques or semiconductor technology have their origins in the fundamental laws of quantum mechanics. These have been of vital importance for the development of modern computers, without which today's internet—or any other globally connected network—would not even be possible. Not as known to the public is the fact that every smartphone, every DVD player, and even every bathroom LED can be seen as an offspring of quantum theory. The high economic importance of quantum technology becomes apparent from the fact that more than a third of the gross national product of an average industrialized country is already today being generated by products based on quantum theory. Research results in recent years and decades give rise to the justified expectation that quantum technologies will hold many more aspects in store. They will find applications in different fields and enable the improvement of existing technical solutions in many areas. They also open up fundamentally new possibilities and perspectives for the future. Besides quantum communication and quantum informatics, quantum sensor technology is of particular interest. Quantum physics is a science of uncertainties and probabilities, yet it has the potential to contribute to unprecedented levels of precision. Even today, classical sensors are becoming smaller and ever more precise. There is, however, not a lot of potential for the technological improvement of the parameters of sensitivity and specificity by classical means. Quantum phenomena such as superposition or entanglement, on the other hand, might be used to record physical quantities such as pressure, temperature, time, position or acceleration, but also electric, magnetic and gravitational fields much more precisely. This is useful for a wide variety of applications, but also for the investigation of fundamental questions in science.

In order to understand how close we are today to this "second quantum revolution", we turn back the wheel of time to take a look at the first quantum revolution. At the end of the nineteenth century, some talented prospective university students (including the young Max Planck!) were advised against studying physics because it was thought there was nothing essential left to discover. However, as Lord Kelvin (William Thomson) put it, "dark clouds" soon appeared in the physics horizon. One of these clouds was, for example, the radiation emitted by glowing bodies, such as the sun, which when at its zenith appears to our human eyes as a brilliant incandescent white. Hardly anyone is aware of the fact that its rays consist of mostly green radiation in the first place. The physiological reason for our perception is that the sun (like every star) emits radiation of many different wavelengths, and human awareness interprets the visible part of this blend as "white" light. From a physics perspective, what is really behind the "green" radiation maximum is Wien's displacement law, which states that as the surface temperature of a star increases, the wavelength of the maximally emitted radiation decreases. A star that is extremely hot, for example, will emit blue light. The next stage, for example our own medium-hot sun, shines a green light. The light of a cool red giant like Betelgeuse in the constellation of Orion is predominantly in the red spectrum.

If we aim to explain not only the maximum radiation, but the entire energy distribution of a glowing body, classical physics won't be able to help us. In classical physics, it is not possible to construct a formula that corresponds to the measured data. According to the predictions of classical physics, the resulting values would become infinitely large ("UV catastrophe"), something that is very far removed from reality. It was the German theorist Max Planck who formulated what was hereafter referred to as Planck's law of radiation by making assumptions that previously had been fundamentally alien to classical physics. Its basic premise is that radiation does not exchange energy in arbitrarily fine gradations. Rather, this energy exchange occurs in discrete lumps or portions, which he called "quanta". For a long time, Planck distrusted his own quantum model. Indeed, he hoped that his assumption would be rejected in favor of classical physics—a hope that was never fulfilled. The concept of quanta turned out to be essential. This insight was provided by Albert Einstein, who at the time was completely unknown and who, inspired by Planck, published his Nobel Prize-winning light-quantum hypothesis in 1905. These light particles, the so-called photons, will also play a decisive role in future quantum technologies. Moving beyond Planck, who initially only assumed the energy exchange between atoms to be quantized, Einstein extended this idea to light, which he said also consists of discrete energy quanta. In this way, the photoelectric effect was interpreted for the first time. Light meters in cameras and photovoltaics applications exploit this phenomenon. Einstein also immediately identified the problems of his new quantum theory, which, as we will see later, finds expression in the double-slit experiment (Sect. 3.1). The resulting interference effect is easy to explain on the basis of the classical wave concept of light; however, it causes fundamental difficulties with the particle conception. With Einstein's interpretation of the photoelectric effect, we would have to predict that for very weak light, i.e. individual light quanta, such interference (casually: superposition of light waves) cannot occur. Yet all experiments prove the exact opposite. This puts our human minds and imaginative capacities to a hard test. For how can a single particle pass through two slits simultaneously—if it is assumed per se to be indivisible? This is against all common sense! These and similar questions became more and more pressing when quantum mechanics was mathematically formulated sometime later by physicists like Werner Heisenberg and Erwin Schrödinger. And this is what we mean when we speak of the first quantum revolution. It enabled us to understand a wide range of phenomena that remain completely unexplainable by classical physics. Not only the radiation of incandescent bodies, but also the creation of light as a quantized transition of atomic spectra (quantum leap) were explained for the first time. First and foremost, however, the modern understanding of atoms and molecules led not only to quantum chemistry, but also to solid-state physics, which provides the direct basis of semiconductor technology, without which modern computers today would not be thinkable. In addition to a multitude of other developments, important applications in medicine (as for example magnetic resonance imaging or positron emission tomography) and, of course, the white LED, which heralded a revolution in luminaires, should also be mentioned. Quantum technology has therefore long since entered our everyday lives.

Despite its enormous accomplishments, quantum physics has the image of being an enigmatic field of science, with confusing fundamental assumptions that challenge the human mind in extreme ways. This perception arises mostly if one is unable to accept quantum physics as fundamental science and attempts to mentally grasp it using the conceptual tools of classical physics. It has been and still is the subject of numerous interpretations and philosophical considerations that remain controversial today. Numerous physicists, however, follow the pragmatic view of Sir Karl Popper. They support a perspective of quantum mechanics that is reasonably free of any interpretation. Loosely based on the motto "Shut up and calculate!", this approach is justified by its continued success, of which there is no end in sight. In the coming years and decades, we might open another chapter in this success story. Many scientists today are not particularly happy with the term "second quantum revolution". True to the motto that one should not count one's chickens before they hatch, the phrase still awaits a more befitting label. Perhaps it would be more appropriate to speak of the "quantum advantage" or of an "advantage through quantum technology". Many of the new applications will not reinvent the world after all, but they may improve existing technical solutions considerably, which in the medium to long term will hold great advantages for research, business and society. The greatest advantage, if not to say technological quantum leap, would be the development of a powerful quantum computer, which would undoubtedly form the culmination of this development.

The long-term goal of networking quantum computers in a tap-proof quantum internet is not restricted to opening their potential to everyone. Such a network would also improve their performance—similar to networks of supercomputers—to unknown levels. Of course, "quantum IT" is still in its infancy. Scientists all over the world are working hard to turn this vision into reality. One major milestone is the development of a regional, and ultimately global, QKD network for quantum cryptography. The renowned Innsbruck quantum physicist Rainer Blatt states with conviction: "Quantum cryptography will be the first economic application of the second quantum revolution". To achieve this, fundamental research is essential. A recent milestone towards the quantum internet indicates this. I have kept the following account in a somewhat journalistic style.

# 1.3    The Quantum Satellite

China, Gobi Desert, Xingjiang Space Base. The date is 16 August 2016, 1:40 a.m. local time.

"…3, 2, 1… We have ignition!" The desert is trembling. The rocket is of the type "Long March 2D". It is many meters in height. It shakes and shudders, wrapped in a white column of smoke. Many people are watching in fascination. One of them follows the spectacle with something like trepidation: Jian-Wei Pan, the Chinese chief scientist of the project. A gentleman of completely different stature is standing by his side. The light of the rocket fire is reflected in his glasses and then lost in the thicket of his abundant beard, giving him the appearance of an eminent philosopher. This is what we might imagine Plato or Aristotle to have looked like—without the suit and tie, of course. This commanding personality is the renowned Austrian physicist Anton Zeilinger.

"Please, nothing must go wrong now!" the two scientists may have thought in unison. But as the rocket ascends smoothly, gaining height effortlessly, the tension diminishes gradually. Inner peace returns slowly. Then, the crowd begins to cheer.

The rocket rises and rises. Already it is starting to cast off its stages. Finally it disappears from view and swings into the earth's orbit. Now the rocket is a rocket no longer. Now it is a 600 kg research satellite, on board the "Quantum Experiment at Space Scale" (QUESS) experiment. Full to the brim with highly sensitive equipment, it will orbit the earth for at least two years at the considerable speed of about 27,000 km/h. The two researchers may send a hurried prayer to the heavens, for the experiments they have planned to actually succeed. They are not worried about the accuracy of quantum physics—of that, they are convinced. Rather, they are uneasy whether the highly complicated technology will function as planned without technical problems. Repairs on the QUESS satellite, that much is clear, would unquestionably pose a few challenges. Already, the pair are surrounded by journalists. Now, patiently, like monks reciting their mantras, they have to repeat again and again what their project is all about.

How did this Austro-Chinese cooperation actually come to be? After all, our scene is set in China. What does little Austria have to do with all this?

There's more than one reason. For one, Austria has brought forth many important quantum physicists, first and foremost Nobel laureates Erwin Schrödinger and Wolfgang Pauli. Austria's current research landscape is impressive as well, with a number of internationally recognized research institutes. The "quantum capital" Vienna should be mentioned as the geographical stronghold of quantum research in Austria. Of particular note here are the outstanding achievements of Anton Zeilinger and his team. Under the nickname of "Mr. Beam", he achieved world fame first and foremost for the world's first realization of quantum teleportation of optical states. This scientific method, which sounds like something out of science fiction, makes information disappear at place A, only to be replicated again at place B. Jian-Wei Pan, the scientific director of the quantum satellite project, had been a member of that particular research group (and one of Zeilinger's doctoral students). The brilliant visionary is a not only a much sought-after interview partner for Chinese national TV, he was also awarded the prestigious Breakthrough Prize. Zeilinger is in no way inferior to him in terms of media presence. After all, he has the become the figurehead of innovative research as potential engine for economy in Austrian politics. The keyword Austria leads us to Europe. Anton Zeilinger would have loved to implement a satellite project with ESA, the European Space Agency, but did not succeed in this. To cooperate with his Chinese colleagues was the only option left to him. The Austrian contribution to the project was mainly the installation of ground stations in Vienna and Graz, where the data transmitted by the satellite is evaluated. These ground stations are actually astronomical observatories, as for example the "Satellite Laser Ranging Station" in Graz-Lustbühel or the "Hedy Lamarr Quantum Communication Telescope" on the roof of the Institute for Quantum Optics and Quantum Information (IQOQI) in Vienna. ESA at least runs one of the optical ground stations, the one on the island holiday paradise of Tenerife. It should be mentioned that the satellite, which is worth roughly 160 million US dollars, was operated and financed exclusively by China.

Surrounded by journalists, the two famed scientists find themselves in the role of quantum priests compelled to give interviews. To some extent, they are pleased by this. Of course, they do want to advertise their projects and promote their prospects for further research funds. On the other hand, they have to keep the technical aspects as straightforward as possible, which in the case of quantum physics, while advisable, is no easy task. In all their composure, their eyes often light up with concealed, yet very real emotion. They passionately believe that quantum physics is the science of the future. And especially in connection with the internet, which has long since become one of the most important lifelines of humanity. The two try to give an easily understandable presentation, loosely following the motto: Tell stories about quantum physics as simply as possible, but not simpler! And what exactly was it that the two explained? Why is this technology so revolutionary? What's the big deal?

QUESS is an international research program in the field of quantum optics. A satellite, named after the Chinese philosopher Micius, who in the fifth century BC discovered that light propagates in a straight line, is operated by the Chinese Academy of Sciences and various ground stations. The University of Vienna and the Austrian Academy of Sciences act as patrons of the European receiving stations. QUESS is a so-called proof-of-concept project. What is investigated is actually the scientific and technological feasibility of the transmission of quantum optical states over long distances. Specifically, this involves the development of inherently tap-proof quantum cryptography and quantum teleportation. In quantum cryptography, an absolutely random quantum key is created for subsequent data transmission via classical encryption methods over the internet. Quantum teleportation, on the other hand, describes the possibility of making quantum information disappear at location A and creating a 100% exact replica at location B, again in an entirely tap-proof way. In both cases, entanglement is the foundation. One of the key objectives is to prove entanglement over a longer distance than before, creating a new record. Another objective is to create completely tap-proof quantum channels between satellites and ground stations by means of entangled photon pairs over thousands of kilometers. Both achieved distances and safety levels represent absolute innovation. Admittedly, "QUESS 1" cannot be seen as the actual technology carrier of the future. For the time being, its communication possibilities are limited. In the absence of sunlight, its functional principle relies on what scientists refer to as infrared interference. If successful, however, further Micius satellites will be launched. This might already form the first prototypes for secure connections between China and Europe in a few years' time. A first global network, the security of which represents a tremendous advance over the previous one, is conceivable as early as 2030 for project manager Pan.

Gradually, the hustle and bustle surrounding the two star scientists fades away a little. While Jian-Wei Pan is still besieged by a crowd of mostly Chinese spectators, who call him "father of the quanta", Zeilinger finally has the time to stroke his beard with some satisfaction. He turns to his company. Suddenly a German journalist pushes his way to the front and starts to pepper him with investigative questions. "Hey, you, that's all Greek to me! This quantum gobbledygook is unintelligible for normal people! Don't you have a more straightforward explanation? And then, why is all that supposed to be a revolution? There are countless possibilities to protect data on the internet, and multitudes of IT specialists who are constantly developing new methods! What do we need quantum technology for at all? Perhaps you just want to make a name for yourself?!"

To answer this fusillade of questions, a universally understandable explanation is needed which, in very simple terms, gets to the heart of the matter. The satellite that just was launched has a special laser on board. A bit like a machine gun, it fires individual projectiles in succession, one after the other (Fig. 1.1). These "projectiles" are light quanta, photons, the smallest units of light. The laser generates blue light, aiming at a special nonlinear crystal. At the target, a pair of infrared photons is produced from each single blue photon. At the speed of light, the two infrared photons move to the ground stations. What comes now is truly surprising. Even though they are spatially far away from each other, each of these photon pairs forms an inseparable unit. They are somehow connected with each other as if by an invisible bond. In certain physical properties, they are closely linked—without wiring, without signal effects, without anything. Just like that! This apparently crazy property of nature is sometimes referred to as the "quantum spook". It has, however, nothing to do with sorcery. Rather, it's a basic property of nature. We refer to it as entanglement. The primary

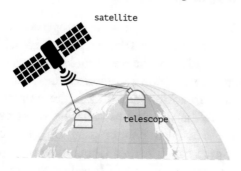

**Fig. 1.1** The QUESS experiment

mission of the QUESS project is to prove that this "quantum spook" actually survives long distances, or that it can be technically maintained. The verification is achieved with the help of statistical methods. As soon as this is accomplished with sufficient scientific significance, the creation of a special key for quantum cryptography can be tested. This can be done for example as follows.

The infrared photon pairs generated by the satellite are entangled in a certain physical property, i.e. they are connected with each other as if by magic. What is interesting, however, is the fact that it remains completely unknown which exact value this property assumes in each case. In order to determine these values, the particles have to be subjected to a measurement. This is done at the ground stations. Again, something very strange happens. Every time such a measurement is done, the value is determined. The result of the next measurement, however, is impossible to predict. It could be the same measurement value, or a different one. In fact, there is no way of predicting the exact value of a single measurement. The result will be completely random, "objectively random" in physics terminology. In particular, such a coincidence cannot be generated by any computer in the world because it arises from nature itself.

If we imagine that the measured properties are arbitrarily assigned the numbers 0 or 1, i.e. one bit of information, the basic idea for the quantum key emerges. Suppose China wants to establish a completely tap-proof data line with Austria. To achieve this (in a future scenario), several satellites could be switched as quantum repeaters in such a way that entangled photon pairs are produced. One photon of each pair is sent to China, the other to Austria. Now, the measurement is performed, with each measurement resulting, completely random, in 0 or 1. Then, the next measurement is done, and so on and so forth. In practice, millions of measurements are performed in a span of seconds, resulting in an absolutely random bit sequence that can be used as the quantum key. This sequence can now be used for cryptographic encoding according to common procedures and sent via the regular internet. The decisive factor is that thanks to the special entanglement-based connection, China and Austria always automatically receive the same key for certain settings of their measuring instruments. This completely eliminates the safety-relevant distribution of keys over the internet. A major step forward!

Why is this technology—if, in fact, it works on a larger scale—tantamount to a paradigm shift in terms of tap-proof data transmission? On the one hand, there is no possibility for code crackers (in practice always supercomputers) to algorithmically reproduce the quantum key, because it originates objectively randomly, i.e. from nature itself. The only option that remains is to sift through all conceivable combinations of bit sequences, which will take an unpredictably long period of time even for supercomputers with typical keys of sufficient length. In addition, each generated quantum key is an original of identical quality that remains independent of cryptographic methods or the trustworthiness of persons. That it is possible to detect any eavesdropping attack due to physics is just as sensational. This had not been possible with earlier technologies. The reason is that potential hacker attacks automatically affect the entire quantum state to such an extent that the resulting statistical record is unambiguous. And so, we always know whether our key distribution is 100% secure or not.

"Uh - all right, professor! It was just a question."

Now that another skeptic seems to be satisfied, even as Zeilinger is not entirely sure whether he has really convinced the figure sneaking away furtively, he now has time to turn to his companion. Heinz Engl, the rector of the University of Vienna, did not want to miss the opportunity to witness this spectacle of radically innovative research from the very beginning. Time for a glass of champagne!

## 1.4 Intercontinental Quantum Telephony

A little over a year later, champagne corks are popping again. The QUESS experiment has proven to be even more successful than expected. The main ceremonial hall at the Austrian Academy of Sciences in Vienna is packed with spectators. Shoulder to shoulder, researchers and journalists are staring at two giant screens. What is about to happen here is a science sensation: the world's first intercontinental video/telephone conference using tap-proof quantum transmission between China and Europe. "This is…", says Anton Zeilinger, who acts as the moderator, "this is not a press conference, but a live demonstration." The second main protagonist is stationed in Beijing, 7600 km east to the east: Jian-Wei Pan, the scientific director of the QUESS experiment. QUESS was started in August 2016 to test the technological feasibility of the innovative quantum communication technology over long distances. Zeilinger lifts a small satellite model into the air, demonstrating the activities about to take place. "Five ground stations in China are receiving satellite data. We are the sixth." The stations are now directly connected via quantum channels. Everyone is waiting in suspense. The minutes are spinning into the void. Only the president of the Chinese Academy of Sciences, Chunli Bai, on the other end of the live video link, is sipping his tea with perfect composure.

To shorten the wait until the connection is established, Zeilinger starts to tell the story of the project's genesis. He criticizes the sluggishness of EU bureaucracy in science funding and praises the swift decision-making processes in China. If his former doctoral student Pan had not asked him to join the project, they would not be assembled here today. Fortunate for Zeilinger, Pan is committed to the Confucian tradition, where a special teacher-student relationship is cultivated. It was Pan who had done the decisive groundwork several months earlier. As early as June 2017, the Chinese researcher and his group succeeded in producing a quantum link over 1200 km for the first time. Already then, that was the absolute world record! This success was an important milestone for quantum communication. It enables two parties to exchange an absolutely secure quantum key over long distances. The security of the key is safeguarded by its completely random generation on the one hand, and on the other hand by the fact that any eavesdropping attack automatically affects the extremely sensitive nonlocal connection of the quanta, attracting attention. When we talk about a nonlocal connection, we mean a connection based on entanglement, the mysterious "quantum spook", which was in this experiment confirmed over a new record distance. Until then, quantum communication had only been demonstrated under optimum conditions over distances of about 100 km. The problem is that both in fiber-optic cables and in atmospheric environments, photons are always scattered by atoms. The consequence is that entanglement is successively lost.

The previous record was held by Rupert Ursin. In 2007, Ursin succeeded in establishing an entanglement-based connection of 144 km between the holiday islands of La Palma and Tenerife. This approaches the maximum distance that can be achieved in this way, which is why the satellite solution was chosen as the next step. In the higher layers of the atmosphere, light quanta are able to move freely, without the noisy effects caused by air atoms. The Micius satellite was launched for this purpose. With an altitude of 500 km, it is only slightly higher than the flight path of the International Space Station ISS. Since satellites close to Earth generally have the highest orbital speeds, Micius is only available to the researchers for a few minutes at a time. This short time span, however, is sufficient to confirm the extremely sensitive entanglement. If the polarization of a light particle at a certain ground station is measured (oscillation plane of the light, Sect. 1.5), the polarization of the entangled partner particle is correlated. Using statistical evaluation methods, for example Bell's inequality (Sect. 1.8), the existence of a quantum channel can be clearly proven.

It is no easy task to explain to non-physicists in simple terms why the researcher's accomplishment is all that groundbreaking. Even under controlled laboratory conditions, such experiments require extraordinary precision. In addition to a special laser and quantum modules, the satellite is also equipped with a high-precision optical system that sends the entangled photons with extreme accuracy to the ground stations. The technical implementation is particularly difficult because it is not possible to simply use standard components. Considerable attention to detail is necessary to render the entire structure suitable for space. In addition to the problem of cosmic rays, which might destroy the highly sensitive devices, even just the control of the optics used for directing the quantum signal to the stations needs to be done with much higher precision than is the case with conventional satellites. Typically, poor precision on satellites is compensated for by higher transmission power. This, however, is not possible with the QUESS project. The quantum signal will arrive only if the entangled light quanta are individually detected in the receiver mirrors, which are 1.2–1.8 m in diameter. At satellite speeds of more than 7.5 km/s, this is a truly tiny mirror size, which makes constant readjustment necessary. For the fast transmission of control signals, the researchers use laser beams with other frequencies, in order not to interfere with the quantum signal. Micius therefore requires high-precision optics that can withstand both violent vibrations (e.g. during launch) and enormous temperature fluctuations. This places very high demands on the design of the experimental setup. Only some of the challenges have been mentioned here, and so it is astonishing how well the coupling between Micius and the ground stations works. In any case, the experiments are a milestone in the development of quantum technologies. To date, China has already realized a backbone with a cable length of about 2000 km between Beijing and Shanghai. However, dozens of intermediate stations (so-called trusted repeaters) are necessary to achieve this, especially since quantum communication in optical fibers without quantum repeaters is only possible over about 100 km. That's why the quantum security chain is constantly interrupted. The QUESS technology, on the other hand, is so revolutionary because it creates a direct quantum channel between very distant points. However, the road to practical applications is still long. Due to problems such as disruptive effects caused by sunlight, measurements can only be taken at night for the time being. Even moonlight can present a problem, which, however, can be compensated for with the help of a sophisticated time mode method. At the moment, however, the real difficulty lies in the data rates, which are still grossly insufficient, even though the special laser generates almost 6 million entangled light quanta per second. Nevertheless, the researchers are optimistic and promise to increase data rates by a factor of one thousand over the next five years.

Meanwhile the hall has fallen silent. It's almost time. Excitement builds. Then, a sudden sound. The signal from Beijing has arrived. "Professor Zeilinger, can you hear us?" The first intercontinental quantum-secured videoconference commences. Thunderous applause. Later, two 100% quantum cryptographically encrypted images are transmitted. The picture sent by Austria is a photo of Nobel laureate Erwin Schrödinger. The picture dispatched by Beijing, on the other hand, bears the effigy of the satellite's patron saint, the philosopher Mozi (Micius in Latinized form). Rupert Ursin explains to the journalists how the new transmission technology works. The entangled photons generated by the satellite are measured by the ground stations, resulting in a sequence of true random numbers caused by the intrinsic quantum randomness. The quantum key generated in this way is then used for a procedure referred to as one-time pad (OTP, Sect. 2.6.1), which in combination with quantum technology demonstrably facilitates completely tap-proof transmission of data via the normal internet. It's important to keep in mind that this is only possible in combination with quantum technologies. On the one hand, this implies that the key distribution must not be done using the internet and, on the other hand, that it is only 100% secure if the used key is absolutely random. Strictly speaking, this condition is not met in normal IT, because classical computers are not able to generate real random numbers. By employing the new quantum technology, however, they are. Of course, Micius is not yet able to establish a direct quantum channel over 7600 km (current record 1203 km). That is why the researchers devised a kind of hybrid system. The satellite sends time-shifted entangled photons to both Europe and China. The light quanta, whose polarizations are in superposition, are first measured by the Graz ground station at the Lustbühel Observatory, resulting in a completely random quantum key, which is stored at Micius. The Chinese then generate a second quantum key in the same way. After that, both keys are mathematically combined in orbit and transmitted back to Austria and China. With their respective keys and the combined key, the private key and the public key, so to speak, both ground stations can generate a common code, which is used for unambiguous encryption or decryption. The special security aspect of tap-proof quantum telephony results from the fact that any eavesdropping attack disturbs the quantum state during the generation of the private key to such an extent that this can be detected metrologically. Unfortunately, due to the limited transmission rate, the generated quantum keys are still too small to securely encrypt the amount of data required for a video call. Therefore, one of them was parsed into many parts and exchanged several times. This means that the phone call was not yet one hundred percent secure.

Nevertheless, Anton Zeilinger is satisfied. "Already…", he says, "we have achieved much higher protection against eavesdropping than anything before." Finally, he again turns to the journalists. "What you have seen here is a historical moment and an important step on the way to a future quantum internet." (https://www.oeaw.ac.at/en/detail/news/pan-jianwei-unter-top-ten-forschern/).

## Japan's First Micro-Quantum Satellite

Quite independent from the Chinese-Austrian work on the QUESS technology, Japan has also taken a pioneering step. In July 2017, the 50 kg microsatellite "Socrates" was launched into space. It can, among other things, be used for quantum communication. On board it carries a mini transmitter weighing no more than 6 kg, which transmits single light quanta with two different polarization directions. These function as 0/1 bits. In contrast to the QUESS method, however, the particles are not entangled with each other. The 10 Mbit/s signal, arriving from an altitude of 600 km, was received by ground stations in Koganei (west of Tokyo), then fed to a quantum receiver, decoded and used for a quantum cryptography protocol. The system developed by the National Institute for Information and Communication Technology (NICT) demonstrates that quantum communication can also be implemented in lightweight and cost-effective microsatellites. This will likely become a key technology for future satellite ground networks, with the long-term goal of a global high-security network based on quantum cryptography. This assessment is supported by the fact that intensive research is already being conducted today on a global satellite communications network with the highest bandwidth. To achieve this, however, it is necessary to develop a technology with the potential to transmit very large amounts of information from space in very short periods of time. Since the RF bands used to date are threatened by overload, the future technology of laser-based data transmission might be a feasible solution. Through the use of laser technology, satellite optical communication functions on a "fixed" frequency band (determined by the frequency of the light) and enables installation in smaller and lighter terminals with greater power and efficiency. In such a purely optical communication, tap-proof quantum cryptography can be implemented directly. (https://www.nature.com/articles/nphoton.2017.107).

## 1.5    Objective Randomness

In the world of casinos and gambling, there's an old saying. "In the end, the bank always wins." This fact (notorious players are only too familiar with) is anything but pure randomness, it is justified by science—by mathematics, to be precise, and even more specifically by probability calculus and statistics. Let's take a look at an example. Take a die, roll it several times and record how many times you get the number 6. Repeat this procedure many times. Make sure your dice are not loaded. Now divide the number of times you have rolled the number 6 by your total number of throws. You will find that the result gets closer to the value $1/6 = 0.1666...$ the more throws you have performed. This is not a coincidence, but a fundamental law of mathematics called the "law of large numbers". For the law of large numbers, it makes no difference at all whether a die is thrown many times in a row or whether many dice are thrown at the same time. What is decisive is only the number of "realizations" of an event, a number which theoretically can converge to infinity. The value $1/6$ obtained in the above example is referred to as the relative frequency of the event. What is meant with the probability of an event is the expected relative frequency of its occurrence for a number of events that converges to infinity. In the above example, the probability of getting the number 6 is about 16.67%. All casino and lottery operators make use of this, because it allows them to calculate their profits with more precision as the number of players increases, and the number of games per player. This is why lottery companies are able to predict their profits with reasonably high accuracy. In modified form, statistical checks are also carried out in economy, by banks and insurance companies; voter transition analyses by opinion pollsters are also performed in this way. In the latter case, a sample is taken, resulting in a conclusive result for a larger collective based on specific distribution functions that indicate the average probability for a given characteristic, including statistically validated confidence intervals.

As can be seen, randomness can usually be managed with the help of probability calculations. But are dice and lottery games (by themselves) really subject to pure randomness? Strictly speaking, this is not the case, because the number we get when rolling our dice, or the balls drawn by a lottery machine, can theoretically be predicted. The exact behavior of dice or balls respectively is predetermined by so-called initial or boundary conditions, as for example exact position, velocity, air resistance, angular momentum, etc. In practice, however, such calculation attempts fail precisely because these initial conditions are not sufficiently known—by a very long shot. One speaks of deterministic chaos or in some cases of chaotic systems. The latter are determined by the fact that the smallest differences in the initial conditions lead to completely different behaviors and are not predictable in the long run. This means that neither the weather nor the exact movements of our planets in the solar system can be predicted over long periods of time. With this, we have found something essential to note. This "principle of randomness" is not real randomness at all, it merely appears as such. In truth, however, there might be a physical explanation for this, which just remains hidden from us. In the sense of cause and effect, there would then also be a causal relationship. Even if a distinction is made between strong and weak causality (in the sense that similar causes have either similar or completely different effects), this possibility cannot be ruled out. The assumption of randomness thus ultimately results from personal ignorance; it is therefore a matter of a *subjective* randomness. This assumption is primarily the basis of classical physics, where real randomness does not exist. If all parameters of an initial state are known (with perfect precision), it is possible to predict the future. This gives rise to the orthodox classical notion that all physical entities, down to the smallest particles, are in permanent interaction with each other. Each action evokes a reaction that is predictable. This scientific approach is referred to as determinism. It was Isaac Newton who was first able to represent the strictly deterministic view in a mathematical way. From Newton's equation of motion (force = mass times acceleration), a differential equation can be formulated for each kinematic process, the solution of which first results in a general set of curves corresponding to the set of all possible states of motion of a body. By specifying the boundary conditions (such as initial position and initial velocity), a concrete (so-called separable or special) solution is created which predetermines the natural process for all times. This results in a description of nature that completely determines the past, present and future of a physical process.

In stark contrast to the deterministic position, however, is the view of randomness in quantum mechanics. Quantum randomness is in fact assumed to be "real". Quantum processes are not subjectively, but objectively random. Thus it is not subjective ignorance which makes them appear to be random. Rather, there simply exist no initial conditions that could be regarded as the cause. This implies no hidden explanation whatsoever can exist. Quantum randomness is irreducible. Today, the overwhelming majority of physicists regard this to be a fundamental property of quantum physics. The coincidence in the quantum mechanical individual process can therefore not be justified further. According to Anton Zeilinger, it is only possible to comprehend why it cannot be explained beyond that point. One example for this is the radioactive decay of a substance. It is commonly described by determining its half-life period, after which half of an original set of atomic nuclei has decayed. This, however, is a purely statistical value. It does not enable deductions at which exact point in time a single atom will decay. Such an event is objectively random. It is, therefore, in no way predetermined by nature whether or not a particular nucleus will disintegrate at a particular point in time. It will either happen, or it won't.

Another example is a photon that passes through a very narrow slit and is then registered on an observation screen. Contrary to what one might think, the light quantum does not pass through the slit in a straight line. Rather, it will be registered at many different points on the observation screen with different probabilities. Again, it is objectively random whether the photon is detected at a certain point or not. In some places it is more likely to be detected, in others it is less likely to be observed. In this case, therefore, quantum theory does not describe the factual, but the possible. The set of all these possibilities (= events that could be measured with a certain probability) is referred to as the quantum state. It is described mathematically by the so-called wave function. In a quantum state, all possibilities exist simultaneously in a kind of superposition until the time of measurement. Only the measurement generates a fact from the existing possibilities. A particle therefore "exists" (until the time of measurement) in all locations where the probability of its measurement does not equal zero. The superposition of all these possibilities is sometimes casually referred to as a "probability wave" (in reference to the key term wave function). Prior to the measurement, the particle is not at any specific location spatially or temporally. This essential property of quantum mechanics is referred to as the *nonlocality* of the world.

## A Quantum Random Number Generator

The special nature of quantum mechanical randomness may seem strange on the one hand, but on the other it opens up promising possibilities for future information technologies. An important requirement concerns the generation of "real" random numbers for tap-proof quantum communication, as demonstrated in the QUESS experiment. The next two sections present two simplified test arrangements.

## Experiment 1

A laser beam hits a beam splitter and is then measured by two photon detectors that are arranged behind it (Fig. 1.2). A beam splitter is an optical component that separates incident light into two beams. 50% of the light intensity is transmitted, the other half is reflected at right angles. A photon detector is a kind of high-precision light meter that can measure individual light quanta using a sophisticated multiplication technique based on extremely sensitive photodiodes. When the laser is turned on, both detectors respond equally—which is not surprising given the 50% split. The arrangement becomes more interesting (and relevant for quantum technology) when a typical laser (which generates light in the form of millions and millions of photons) is replaced by a special single photon laser. This single photon laser enables the generation of light of the "smallest possible light intensity". Like an ancient musket, it fires only single light quanta, one after

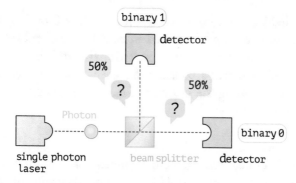

**Fig. 1.2** Quantum random number generator

the other. At this point it should be noted that the brightness (light intensity) in the quantum model corresponds to the number of photons. The smallest amount of light is therefore exactly 1 light quantum. Below that, there is nothing but zero, not a single light quantum, and therefore complete darkness. In contrast to classical physics, which permits an infinite number of gradations between 0 and 1, quantum physics follows the motto "Nature does make jumps" (something Max Planck initially did not want to believe). If the single photon laser sends single light quanta onto the beam splitter, it is observed that one detector responds to the other in a completely random sequence—but never both at the same time. If the binary numbers 0 or 1 are assigned to the two photon detectors and all individual measurement values are arranged one after the other, a bit sequence as for example 0101011001… is created in this way. Now the experiment is repeated, and individual measurements are performed again. The resulting bit sequence is 1100101011… This is clearly different from the first experiment. What follows is a long series of measurements with billions of individual photons. If one then determines the relative frequency, i.e. divides the number of all binary values 0 or 1 by the total number of all measured values, then one will receive 50% for each case as an estimate for the probability. This probability remains constant for each series of measurements (assuming the number of readings supports the law of large numbers).

**Interpretation**

The described arrangement represents a quantum random number generator. When a single quantum of light hits the beam splitter, objective quantum randomness makes it completely impossible to predict whether it will be transmitted or reflected (and thus measured as 0 or 1). Even though the overall statistics exhibit a 50% distribution each time, the individual quantum mechanical event remains unpredictably random. In classical physics or with a "normal" laser, this effect remains hidden only because visible light consists of an astronomical number of photons. Apart from that, the response of only one detector at a time indicates that a photon must be an indivisible physical object. Of course, this statement by itself would not be sufficient scientific proof. Photons, however, are indivisible elementary building blocks of light (and also of all other radiation on the electromagnetic spectrum), as is shown by Einstein's interpretation of the photoelectric effect, but also phenomena that were discovered later, as for example the Compton effect and countless other experiments (including several performed in accelerator facilities such as CERN in Geneva).

**Experiment 2**

In this variation of the experiment, our conventional beam splitter is replaced by a special, so-called polarizing beam splitter. In contrast to the above, the division ratio here depends on the polarization of the incoming light beam. At this point, we briefly explain the polarization of light in both classical and quantum mechanical terms. The Scottish theorist James Clerk Maxwell described light in classical physics as an electromagnetic wave (Heinrich Hertz succeeded in proving this experimentally). If we consider solely the vibration component of the electric field, we find that it can describe different directions, which may also change over time. Linearly polarized light is a special case. Here, the electric field strength always oscillates in the same plane. In view of the particle-like quantum model of light, such an idea seems grotesque. It has led to the somewhat problematic term wave-particle dualism of quantum objects. We will come back to this later. For the time being, we will simply accept the fact that particles can also be assigned the wave-like property of polarization. What happens now when individual light quanta hit the polarizing beam splitter, which in contrast to its typical counterpart also takes the polarization direction into account? Assume that the polarizing beam splitter allows horizontally polarized light to pass in the transmission direction and that a photon detector with the inscription "0" is placed behind it. In the reflection direction, on the other hand, only vertically polarized light can pass through. Behind it, detector "1" is mounted. Now we assume that the light emitted by the single photon laser is linearly polarized. The experiment can provide three results. In the case of horizontally polarized light, detector 0 will always respond with certainty, i.e. the probability of 100%. In the case of vertically polarized light, only detector 1 will "click". Things are very different if the polarization of the incident light lies between the horizontal and vertical directions, for example at an angle of 45° to the beam splitter. Then, objectively and randomly, first one detector will respond, then the other, and a quantum random number generator will be created (as in experiment 1).

This quantum random behavior also applies to other angles, for example 30° or 60°, but with different overall statistics and thus different probabilities (see "Malus's law" in Sect. 2.6.3).

**Interpretation**

Obviously, the described arrangement can also act as a quantum random number generator when a polarizing beam splitter is used (as long as the incident light is neither horizontally nor vertically polarized). In these relevant cases, the behavior of the photon is of particular quantum mechanical interest. After the light quantum has passed through the beam splitter, it is in a superposition of two possibilities, namely transmitted or reflected. The abstract "probability wave" represents this quantum state, which, however, can by no means be understood as a concrete, spatially extended wave. Its task is simply to indicate the probability of which events may occur. All existing possibilities are already contained in the quantum state, at the same time. However, since a photon is indivisible, once it has been measured at one of the two detectors, it cannot be registered at the second detector as well. The superposition therefore necessarily has to collapse in the moment of measurement (since it describes the probability, which after the measurement has to be zero at the second detector). In the spirit of the Copenhagen interpretation of quantum mechanics (named after the Danish physicist Niels Bohr), we speak of a "collapse of the wave function". That the superposition principle, which might seem somewhat unsettling, is definitely not the product of an overactive imagination, is proven by the fact that the detectors can be removed in the experiment and the partial beams reunited in a second beam splitter. The result is a photon with the original polarization direction.

# 1.6  Quantum Entanglement

As a central element of quantum mechanics, quantum entanglement is one of the most interesting phenomena in physics. It is also an essential resource for a future quantum internet. It enables inherently secure quantum key exchange between communication partners, and also provides crucial tools for the realization and networking of quantum processors.

**Perfect Correlations**

In Sect. 1.5 we have seen that the combination of single photon laser, polarizing beam splitter and photon detectors results in a quantum random number generator. Now we will take a look at an arrangement with *two* such random number generators. In particular, two polarizing beam splitters are used, which can be widely separated spatially. An entanglement source (e.g. specially excited calcium atoms) is located exactly in the center. It emits a photon pair, each photon in the opposite direction, which is correlated in polarization. This means that the emitted photon pairs are very closely related in certain physical properties—in this case, the polarization of light. Usually, the expression is used that the polarizations of the photons are "entangled". If we now perform a series of measurements on one of the two beam splitters, this results in an objectively random sequence of binary numbers, in analogy to the two experiments in Sect. 1.5. What is most astonishing about the result is the fact that the parallel measurement series at the second beam splitter delivers exactly the same binary sequence! For a new measurement series, the resulting bit sequence is different, but the measurement at the second generator results in the exact same bit sequence as at the first random number generator. This experiment can be reproduced in any sequence and the same type of perfect correlation is found in every single case (without the possibility of measurement errors).

## Interpretation

In view of the fact that the binary sequences are objectively random, the behavior of entangled systems is highly remarkable. Since it is completely random whether the number 0 or the number 1 is measured at one of the two random number generators, it is all the more surprising that the measurement at the second generator always results in exactly the same sequence of binary numbers as the first measurement. This clearly proves that certain observables of entangled particles cannot be statistically independent but rather, they must be strongly correlated. This statement challenges the human mind to the extreme, especially in view of the fact that entanglement has already been verified by QUESS over a distance of more than 1200 km. We might therefore separate the two random number generators at a distance of 1200 km from each other, and they would still exhibit the perfectly correlated behavior described above. As quantum cosmology suggests, this seemingly bizarre phenomenon seems to be valid throughout our entire universe and might even present a general rule.

## Anticorrelations

In the example outlined above, perfect correlations occur, i.e. the polarization of the two entangled particles is exactly the same during their measurement. There exist, however, also entangled systems that are anticorrelated. For example, photons can be entangled in such a way that the measurement on one of the two particles produces a vertical polarization, whereas the parallel measurement on the entangled partner particle produces a horizontal polarization. The two polarization planes are thus rotated

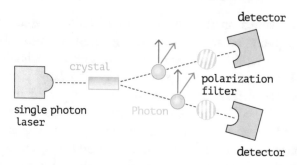

**Fig. 1.3** Generation and measurement of anticorrelated photons

by 90° against each other. A corresponding experimental arrangement is shown in Fig. 1.3. A single photon laser targets a special crystal, creating a pair of entangled photons from each single quantum of light. Polarization filters, which act as analyzers, are used to measure the anticorrelated behavior of the generated photons. Statistically, we will receive 100% of the "detector clicks" at both detectors only in those cases where the position of the analyzers

1. corresponds exactly to the polarization direction of the generated particles and
2. the relative position of both analyzers to each other is exactly 90°.

It is subject to objective randomness which polarization is determined at a single light particle. As soon as the measurement value is determined, however, the second measurement value of the partner particle is automatically predetermined as a result of the entanglement. In a certain sense, quantum physics in this case behaves deterministically. While measurement values are subject to objective randomness, the further course of events is predetermined. Nevertheless, the phenomenon of entanglement cannot be explained on the basis of classical physics, which fundamentally separates quantum mechanics from classical theories. On the other hand, quantum mechanics is in harmony with the special theory of relativity, which (because of its causal structure) is also regarded to be a classical theory. This results in important conclusions for the safety aspect and the transmission speed of a quantum internet, which will be discussed in detail later (Sect. 3.6).

Another example of anticorrelated behavior is the quantum mechanical property of the spin of a particle. Spin can be imagined as a kind of inner twist (angular momentum) of a particle. It occurs in both bosons (e.g. to photons and corresponding to their circular polarization) and fermions (such as electrons or protons) but can have different values. For bosons, the spin is always an integer (photons for example have spin 1), for fermions it is always a half integer (electrons have spin 1/2). Like many physical quantities, spin can also be assigned a direction with respect to a given axis. Thus, the spin of an electron can point in a positive ("spin up") or a negative direction ("spin down"), i.e. it can amount to +1/2 or −1/2. These numerical values refer to the proportionality to the quantum of action. This is because naturally, spin is "quantized" as well. In addition, for spin, there exists a principle of indeterminacy that is comparable to the Heisenberg uncertainty relation. This implies that the elements of the spin cannot be measured simultaneously in two spatial directions. The spins of quantum particles can also be entangled experimentally, exhibiting anticorrelated behavior. An entanglement source, for example an atom, emits two spin-1/2-particles which move in opposite directions from each other. If we then measure "spin up" on a particle, we automatically measure "spin down" on the entangled partner particle, and vice versa. If "spin down" is measured on a particle, then the entangled particle certainly has "spin up". This anticorrelation holds for every direction. It also applies, for example, to the horizontal or vertical direction. If we therefore measure "spin right" in the horizontal direction, the entanglement at the second particle delivers "spin left". A measurement in the third direction of space also behaves in the same way. For example, the measurement result "spin front" at the first particle delivers a "spin back" result at the second particle. As explained above, concrete measurement values are always subject to objective randomness.

**Entangled Many-Particle Systems**

Entanglement is typically represented as a phenomenon of two closely correlated particles. We might get the impression that this phenomenon is limited to pairs of objects. This is by no means the case. Relevant phenomena such as GHZ states (GHZ = Greenberger/Horne/Zeilinger) and related experiments have proven that several or even large numbers of particles can be entangled with each other, even if they are spatially separated. Complex entangled systems are an important area of research, particularly in view of expected future developments in quantum information technology. A rule of thumb is that the conditions get ever more useful as their complexity increases, which on the other hand makes it much more difficult to deal with them. One possibility for concrete realizations is to allow a large number of particles to interact with each other in order to create complex entanglement states in this way. The objective of this research is not only to gain deeper insights into quantum mechanics, but also to investigate further fundamental questions. Of technological importance, for example, is the question of whether the amount of information which entangled quantum objects can "store" is unlimited or whether we will reach a fundamental limit. However, the realization of entangled multi-particle systems is a very difficult task and presents quantum technology with enormous challenges. Already systems with more than three particles prove to be very difficult to manage. As current research results prove, the development course is on the right track. In many-particle systems, entanglement is defined by the so-called entanglement spectrum. This makes it possible to record important properties of a collective quantum system, including how difficult it is to calculate them using classical computers. Instead of using a quantum simulator to measure the entanglement properties of the realized state, the entanglement operator is simulated directly. This method offers the great advantage that the spectrum of much larger quantum systems can be measured, which would be very difficult, if not impossible using conventional computers.

## Quantum Mechanical Interpretation

Although initially "trivialized" as a purely statistical correlation and later ridiculed by Albert Einstein, entanglement proves to be the distinguishing element of quantum mechanics par excellence. This statement was already made around 1935 by Erwin Schrödinger, who also coined the term "entanglement". Its main feature is that the components of entangled systems are not localized individually. Rather, their *mutual* condition is spatially distributed over the entire system. This phenomenon can therefore only be described correctly on the basis of a nonlocal theory. Quantum mechanics explains entanglement on the basis of the superposition principle outlined above, which also refers to states of composite systems. Only in the special case where the overall state is equal to the product of its individual states are subsystems independent of each other. In general, however, they are not, and therefore entangled. Consequently, they can only be described by a single state that represents the entire system. In other words, an entangled state is to be seen as an abstract entity that can expand independently of space and time, theoretically over arbitrary distances (i.e. is not local) and can definitely not be traced back to separate subsystems. Only with the help of these nonlocal correlations can a quantum-mechanically complete description of the overall system be obtained. Although the mainstream of science follows this interpretation, even today there remain a few "incorrigible relativists" (quote Rupert Ursin in conversation with the author) in desperate pursuit for a classical theory, with occasionally alarming explanations. Nature, however, seems to follow the well-known slogan: "The whole is greater than the sum of its parts."

## 1.7     Spooky Action at a Distance

In popular literature and in media reports, entanglement is sometimes referred to as "quantum spook". How does this mystifying term actually come about? Is it just a journalistic commonplace? Not in this case. The term can be traced back to someone you will certainly be familiar with: Albert Einstein, probably the world's most popular physicist along with Stephen Hawking.

Albert Einstein, who is known for the genius of his world-famous theory of relativity and his pacifist-liberal convictions as well as for his comically eccentric appearance, which was once compared to that of a retired shepherd dog. Albert Einstein, who occasionally has to serve as a metaphorical anchor of hope for poor students, because he was repeatedly (and erroneously) portrayed as one of them. It certainly won't be necessary to explicitly declare that only a handful of people will go on to discover laws of nature. In the case of theoretical physics, this obviously requires an extraordinary talent for mathematics and complex logical thinking. Another unique feature of a research genius is a powerful instinct for physics. This instinct was particularly strong and rich in Einstein's case.

Albert Einstein (1879–1955) was born as the son of a Jewish family in Ulm. A German native, he later acquired several other nationalities. For a short while, he also was an Austrian. Described as a bright, occasionally seditious, yet very talented student with a keen interest in the natural sciences, he was during his studies at the Swiss Polytechnic depicted by his mathematics teacher as a "lazybones", conspicuous through his absence. Later, when formulating his general theory of relativity, Einstein would come to regret this earlier attitude. Einstein, as a theoretical physicist, was naturally very interested in mathematics—but only as an aid for describing physical models. He was most skeptical about mathematics per se as a highly abstract and rather exotic subject (mathematics by its very nature truly is very abstract, of course). Einstein was all the more grateful later when he was able to build on significant elementary work and the support of the mathematician Hermann Minkowski and his friend Marcel Grossmann. His famous miracle year ("annus mirabilis") 1905 has become legendary. This was the year that Einstein submitted his photon hypothesis and his special theory of relativity, both of which were genuinely revolutionary at the time. On the basis of the Brownian molecular motion, Einstein also provided evidence for the existence of atoms (the reality of which was still highly controversial at the time). At the time, he still was a scientific nobody who had applied in vain for the position of University Assistant and, instead frittering away his time as an employee of a Swiss patent office as "3rd class assistant". His works in physics were so innovative that they were met with nothing more than disapproving head shakes at the time. Today they are celebrated as the discoveries of the century, comparable those of Copernicus, who also radically changed the way we see the world. When, in 1915, Einstein's general theory of relativity with its famed concept of space curvature (which was verified experimentally shortly later) was published, the recognition of Einstein's work increased beyond all borders. He turned into the world's most popular physicist and has become iconic for the ultimate research genius. Even today, however, it is still considered a scandal that Einstein was awarded the Nobel Prize only once (in 1921 for his photon hypothesis), despite the fact that with his theory of relativity, he practically developed an entire branch of modern physics all by himself. The reasons are open to speculation. Was it because of the incompetence of the Nobel Committee or because of his Jewish ancestry? Or was it because his theory of relativity was too daring for the times?

Immediately after the Nazis seized power, Einstein emigrated to the USA, where he received a visiting professorship in Princeton which he held until his death. As one of the world's most influential thinkers and as a pacifist, he was also in great demand as a consultant for political and military matters. He later expressed that his recommendation to President Roosevelt that the USA launch a program to develop an atomic bomb was a "big mistake". His motivation was the concern that Hitler's Germany under Werner Heisenberg and others might develop such a weapon of mass destruction before anyone else could. At no time, however, was Einstein himself actively involved in the "Manhattan Project". Einstein's ambivalent relationship to quantum physics is remarkable: on the one hand he was an important pioneer (though not creator), on the other he did not accept its epistemological consequences. Because of this, the genius increasingly squandered his "intellectual leadership" role in theoretical physics in his later years, as one colleague once put it. While he was celebrated in public like a popular hero, the quantum community found their former champion to lag behind the times. Why was Einstein so reluctant to accept the consequences of quantum theory?

### The Marble Game Analogy

Two people, let's call them Alice and Bob, live in two different places. One day, they decide to play a special game. Each of them mails the other a package containing a certain number of red and blue marbles. The rules of the game stipulate that both of them have to open their packages at exactly the same time. Another rule states that they have to wear blindfolds while doing this. After precisely one marble has been removed from the package, it is closed again without the receiver ever looking inside, and thrown away. This is repeated every three days and goes on for a long while. Over time, Alice and Bob notice something strange. With every round, both of them draw marbles of the same color. Whenever Alice picks a red marble, Bob extracts a red one from his package. If Alice's marble is blue, then Bob's marble is blue as well. And yet, the colors appear to emerge randomly. Alice and Bob, perplexed, wonder what the reason could be.

Enter two theorists. Let's call them Niels and Albert. They are also looking to explain the phenomenon. Niels: "I conclude that both packages form an inseparable unit with each other, even if they are spatially distant. The color of the marbles is fundamentally indefinite. The marbles only take on a concrete color when they are drawn by Alice and/or Bob. But whether they are red, or blue, is completely random. Objectively random. All that is possible is to make a statement about the probability." Albert immediately protests: "What nonsense! There are several possible causes. For example, the post office always only places marbles of the same color in the packages. Or it was an offender who, in the night, manipulated the packages. Alice and Bob could've made a deal somehow. We just don't know it. I conclude that it is simply our ignorance that makes the appearance of the colors seem to be random. Their probability assumption thus results from personal ignorance and cannot be a law of nature."

Their heated debate goes on for a while. After some time, another theorist joins them: John. He claims he can propose an experiment that clearly proves which of the two is right.

What will John's mysterious measurement reveal? Of course, you would guess that Albert is right. Niels' explanations seems to be rather far-fetched and diametrically opposed to our everyday experience.

In fact, our "marble game" is only a caricature of one of the most quoted debates in the history of science, the famous controversy between Albert Einstein and the Danish Nobel laureate Niels Bohr about the so-called Einstein-Podolsky-Rosen (EPR) problem. We remember that today, the correlated properties of entangled objects are interpreted on the basis of nonlocal quantum theory. It is precisely this nonlocal aspect that puts the human mind to an even harder test. Human logic automatically assumes some cause "behind" it, following the causal approach, which is firmly embedded in the human imagination. Already at the outset of quantum theory, two opposing schools of thought emerged. The followers of classical realism were led by Einstein (but also Erwin Schrödinger, Louis de Broglie, David Bohm and John Bell. Their opposition are the representatives of a more positivist philosophy, led by Niels Bohr, Werner Heisenberg, Paul Dirac and Wolfgang Pauli. Niels Bohr was familiar with Sören Kierkegaard's philosophy and Immanuel Kant's investigations into the "thing-in-itself". The latter had concluded that nothing can be said about the thing in itself, not even whether it exists at all. Our messages can therefore only concern experiences and perceptions, observations and measurements. This is also Bohr's argument in the so-called "Copenhagen interpretation of quantum mechanics". Only what is observable can be known about the world. Physics should therefore deal with observable quantities (observables). In general, verifiable statements are only possible via probabilities. Einstein, on the other hand, was (in the jargon of philosophers) a representative of "naïve realism". It was his conviction that things exist even without observation, with precisely defined properties. He refused to accept the concept of quantum randomness ("*God does not play dice!*") Instead, he advocated an objective, complete and precisely determined reality. While Bohr postulated the principle of complementarity, Einstein considered Heisenberg's uncertainty principle (which expresses the complementarity of two physical quantities by describing the impossibility of measuring place and momentum simultaneously) to be merely a shortcoming of measurement accuracy. Around 1930, several meetings were held for the presentation and discussion of new developments in quantum physics, including the famous debates, which were acutely intensified by the example of quantum entanglement.

Let us have another look at quantum entanglement and consider what makes it so poignant. As our example, we take a system of anticorrelated spins, similar to what we learned earlier. Spin up is measured at particle A. Shortly afterwards, spin down is measured at particle B. This result in itself would not be of particular interest, because the measurement results might have already been determined during the generation of the particles. The explosive issue is the fact that the measured value at particle A is objectively random. It could just as well result in a spin down of particle A and thus a spin up of particle B. Or a corresponding spin correlation in any other spatial direction. The special thing about quantum entanglement is that *despite* the objective randomness of the measured value of particle A, the measured value of particle B is always predictable. In other words: the result of measurement value B depends on which measurement value is obtained at A. In modern experiments, it is also possible to measure A and B faster than the time light would need to travel from A to B. The measurement of A and B can be performed in the same way as in modern experiments. In this sense, the wording that that A and B influence each other instantaneously is justified.

Now we put ourselves in Einstein's position and formulate our explanation along the lines of the "realists". Essentially, there are two possibilities:

1. A (still) undiscovered mechanism might exist, which determines the behavior of the entangled particles in advance (and thus also simulates objective randomness). This is consistent with the spirit of determinism as it was so successfully introduced into classical physics by Newton, among others.

2. If objective randomness were indeed a law of nature, the entangled objects would inevitably have to somehow communicate with each other, i.e. exchange classical information. This has to be true because the entangled second particle has to "know" which measured value is present in the first particle in order to "show" the anticorrelated spin. If the two particles are far removed spatially, the question immediately arises as to how fast this signal effect can occur. Is there a limit?

Long before the corresponding experiments could be realized (let alone milestones like QUESS), Einstein repeatedly demonstrated his instinct for physics and his ingenuity. For example, he was the first to grasp not only the fundamental significance of Planck's quantum model, but also the particularly explosive nature of entanglement, which he was familiar with from theoretical considerations. While elsewhere entanglement was still played down as a statistical correlation, it was already clear to the great thinker that it had to be a phenomenon of fundamental relevance. Although the genius was deprived of the actual fruit of his achievement because he drew the wrong conclusions from it, he made a decisive contribution to the fact that this (originally purely philosophical) question was even addressed in physics. It is all too easy to understand that Einstein naturally trusted "common sense" in quantum theory at the time of its formation and was looking for an explanation in the above sense. In the case of entanglement, Einstein followed intuitive human logic and questioned not only the uncertainty relation, which he regarded as a shortcoming of measurement accuracy, but also objective randomness. This mindset is associated, among other things, with the often-quoted words "*God does not play dice!*" According to Einstein, there had to be a cause for the phenomenon of entanglement. Physics, Einstein thought, simply had not yet discovered those "hidden variables", as he called them. He argued along the following lines. If it really were completely random which features objects exhibit upon measurement, then one particle would have to send a message to the other, send a signal, exchange classical information. Such classical information transfer would have to occur with superluminal velocity, given the considerable spatial distance between the two objects. Of course, Einstein viewed this to be in clear contradiction to his own theory of relativity, which expressly prohibits the transmission of information at superluminal velocities. This also contradicts the relativistic principle of locality. If such a thing were indeed possible, then, he thought, something's not quite right. Einstein spoke of "*spooky action at a distance*" between the entangled objects. Of course, this simile is to be read in a mocking sense. It can be seen as a veiled request directed at his colleagues to further develop quantum mechanics into a complete theory. According to Einstein, quantum mechanics would be either nonlocal or incomplete. Since he was not able to get used to the nonlocality of entangled systems, he concluded that quantum mechanics had to be incomplete.

Now we know where the expression "quantum spook" originated. Einstein later added a philosophical argument to the discussion. He asked about the reality of physical objects. In particular he searched for physical entities which he referred to as "elements of reality". If these objects actually only become reality when they are measured, then they cannot exist prior to that. In any case, Einstein defined a physical quantity whose value can be predicted with certainty without disturbing the system as an element of reality. In a complete theory, each element must have its counterpart in physical reality. Not to create the wrong impression, Einstein did see quantum theory as "compelling"; after all, he himself had been involved in its development. He had no other choice, because the formalism (to this day) is in best agreement with experiment. He just had immense difficulties with the epistemological or ontological side, which in his opinion forms an essential foundation of any physics theory. Once, he harshly snapped at the young Werner Heisenberg: "You, if you think you are able to develop a theory of observable quantities, you are grossly mistaken!" By this, Einstein meant that only the philosophical corset of a theory would decide what the observable variables would be—and that framework, in his opinion, would be local realism.

Just how important local realism was to Einstein can be seen from the fact that he enlisted the support of the young American physicists Boris Podolsky and Nathan Rosen (probably also for linguistic reasons). Einstein, Podolsky and Rosen (EPR), in an article published in the USA in 1935, raised the question of whether quantum mechanics was a complete theory (Einstein et al. 1935). Although Einstein was not all that fond of this work, because its essence was "buried in erudition, so to speak", he worded that essence in more combative terms in his mother tongue later, when "Quantenmechanik und Wirklichkeit" was published:

> It also seems essential for this classification of things in the framework of physics that at certain times, these things exist independent from each other, insofar these things are located in different areas of space. Without the assumption of the independent existence (essence[1]) of things that are spatially removed from each other, which first and foremost has its roots in our everyday thinking, thought in terms of physics as we know it would not be possible. (Einstein 1944)

This position, which today is referred to as local realism stands in clear contradiction to the current view of nonlocality in quantum theory. For the work in question, EPR therefore set out a criterion for reality and locality that can be represented in a simplified way. Two entangled spin-$1/2$-particles are emitted from an entanglement source. According to EPR, if the particles are sufficiently far away from each other, it should be possible to perform a measurement on the first particle without affecting the second particle. Since they are not able to influence each other, all spin values that could possibly be measured on the particles have to be determined in advance. As shown above, one would then be able to measure "spin up" on one particle and "spin down" on the other. This has to apply also in different spatial directions. The quantum mechanical

---

[1]Translators note: The term „So-Sein"in the German original was translated with the word „essence".

uncertainty relation, however, does not permit the simultaneous measurement of spin components of different directions. According to EPR, this would result in a contradiction, from which it follows that quantum mechanics must be incomplete. It would therefore be necessary to replace quantum mechanics with a more fundamental theory, which enables the simultaneous calculation of all spin components, as well as the simultaneous determination of location and momentum (rendering Heisenberg's uncertainty relation obsolete). At Einstein's time, modern experimental physics had not progressed so far that it might have led to a decision. And so, the genius indulged in endless discussions with the Danish physicist Niels Bohr on a purely theoretical level. In the famed Solvay Conferences which took place during the 1930s, Einstein incessantly sought to outwit quantum mechanics. The congenial Niels Bohr would always be able to parry the intellectual blows by his colleague. In one of those cases, Bohr seemed to almost have arrived at a checkmate. But then, he conjured Einstein's general theory of relativity out of thin air and at the last moment beat him (in league with Heisenberg and Pauli) with his own weapons.

## 1.8 Bell's Theorem

As outlined above, Einstein, Podolsky and Rosen (EPR) in their 1935 work raise the question whether quantum mechanics, as a theory, is actually complete. This argument is often referred to as the EPR paradox. At its heart is the demand to complement quantum mechanics with a theory of hidden variables, because otherwise, entanglement would appear paradoxical. Most notably, according to Einstein, something like "spooky action at a distance" occurs. And that really is not a suitable approach for exact natural sciences. Yet, paradoxes only arise when classical patterns of thought are applied to quantum mechanics, such as local realism. As we know today, no physical theory of hidden variables is capable of reproducing quantum physics in all its predictions. The so-called Bell theorem is of essential importance, since it shows that this bizarre quantum world "really" does exist, and that its description of nature has to be of fundamental importance. The fact that it is based entirely on scientific evidence is a crucial factor in substantiating this finding. For the general reader, however, the question may arise how such a demonstration of evidence is actually done, in concrete terms. To illustrate this, a short introduction into the basic methods of physics research will follow.

### Falsifiable Hypotheses

We start with a straightforward example, which is easily recreated at home. Take a sheet of paper and a coin. From the same height and at the same time, drop both objects, so that they fall to the ground. You will observe that the coin falls much faster. Why is that? Essentially, two reasons are conceivable.

1. The coin falls faster because it is heavier, i.e. because its mass is greater;
2. the coin falls faster because the sheet of paper has more air resistance and a force (friction) is at work which slows down the speed of the fall.

We have now formulated two assumptions (hypotheses) which can be either true or false. This already meets an important scientific requirement: the formulation of *falsifiable* hypotheses.

In order to determine which of the two hypotheses is true and which is false, a suitable experiment that leads to an unequivocal decision needs to be performed. In the words of Isaac Newton, the experiment is the "highest judge" in physics. To proceed with our experiment, crumple the sheet of paper into a ball and again drop both bodies to the ground from the same height. You will find that both objects fall at almost the same speed. In this way, hypothesis 1 was automatically disproven, hypothesis 2 on the other hand was confirmed. This is because of a very simple logical conclusion. Hypothesis 1 assumes that heavy bodies fall with higher velocities. That means that the coin should again reach the floor earlier than the sheet of paper, whose mass has not changed as a result of the crumpling. The outcome of the experiment contradicts hypothesis 1, which therefore must be rejected. Hypothesis 2, on the other hand, is clearly confirmed, since a reduction in air resistance also reduces the resulting friction. And this is why the paper ball falls more rapidly.

Based on this experiment, we can now develop a further hypothesis. If free fall obviously does not depend on mass, but only on air resistance, then all bodies in a vacuum must necessarily fall at the same speed (since there is no air in a vacuum). To experimentally test this hypothesis, our budget needs to be a little more generous. We can, for example, use a long evacuable glass tube in which we place a downy feather and a coin. With the help of a reasonably efficient vacuum pump, it can easily be demonstrated that both objects clearly fall to the ground with equal velocities. And this is true in spite of the fact that the mass of a downy feather is certainly much smaller than that of a coin.

To summarize, it remains to be noted that scientific evidence is produced in the following way.

1. formulation of falsifiable hypotheses
2. verification of these hypotheses via reproducible experiments

The verification of a first hypothesis can result in conclusions that lead to further experimental tests. Very often, one will in the course of scientific research come across an interesting effect. An appropriate hypothesis is frequently formulated to explain the effect. In such cases, one speaks of an interpretation. Einstein's photon hypothesis is a famous example for this. In any case, there's a golden rule that always applies. Only hypotheses that have been confirmed by experiment have the potential of being recognized as laws of nature. In simpler terms, physics always acts according to the maxim: "Assertions have to be proven *conclusively* (one would also like to suggest this approach to politicians)." These rules of the game also apply to Bell's theorem in a much more complicated way. Here, too, suitable hypotheses have to be formulated, which then must be tested by means of experiment. Such assumptions are usually formulated in mathematical terms. Based on the resulting equations, concrete values of measured variables can be predicted. These values are then compared with the experimentally obtained data. Only if the results acquired hereby are of scientific significance (= the deviation is sufficiently small according to statistical guidelines) do such assumptions have the chance of being considered laws of nature. In the case of the EPR paradox, however, this did not happen until 1964, when a corresponding mathematical criterion was published (earlier experiments were not sufficiently general). The result was what is today known as Bell's inequality. This quantitative representation of the EPR problem is due to an Irish theorist who sadly died far too soon, and to whom the foundations of quantum theory were of particular concern.

### Doctor Bertlmann's Socks

Let's go back to the marble game analogy. Remember John?

John joined Niels and Albert and claimed he'd be able to bring about a decision.

The John in question is actually the Irish physicist John Stewart Bell who, decades after the Solvay Conferences, joined the debate and articulated the EPR paradox into a quantitative formalism. Even though he came from a poor background, Bell obtained degrees in mathematical and experimental physics. Ultimately, he was employed as particle physicist and field theorist at CERN in Geneva. He was also deeply interested in fundamental questions of quantum theory, in particular the EPR argument. Bell once wrote a famous analogy ("Doctor Bertlmann's socks"), in which he summarized the topic in a humorous way that made it easy for the general public to comprehend. The scenario focuses on Bell's friend and colleague Reinhold Bertlmann.

The renowned Viennese quantum physicist Reinhold Bertlmann has (as he himself is fond of repeating) the habit of wearing socks of different colors. If one sock is red, the other one might be blue. If one sock is pink, the other one will probably be green, and so on. According to John Bell, whenever you meet Reinhold Bertlmann, you can be sure that his socks are "anticorrelated", that is, if you observe the color of one sock, then the other one will certainly be of a different color, for example anti-red, anti-blue, etc. If one observes a sock peeking out from underneath Bertlmann's trouser leg (whatever color it is then), it is determined with certainty that the color of his other sock will be different (Fig. 1.4). This is strikingly evocative of the anticorrelations in entangled particles, which exhibit similar behavior. In his humorous metaphorical scenario, John Bell concisely summarizes the core question of the EPR problem. Is the color of both socks predetermined by nature (position of realism) or does the color of the socks come into being by objectively random chance only during the measuring process (observation) of

**Fig. 1.4** Dr Bertlmann's socks (inspired by a drawing by John Bell, who is depicted on the left)

the socks themselves? In the case of Doctor Bertlmann, we can assume a realistic position (which was the position Bell actually advocated), namely that there is a definite cause for the socks' anticorrelated behavior. Surely, Reinhold Bertlmann chose socks of two different colors from his sock drawer in the morning and put them on his feet before stepping into his trousers. If, on the other hand, the socks were real quantum socks, the color of the socks would be completely indefinite until they were observed.

There are the two main positions between which Bell wanted to bring about a scientifically sound decision.

### Hypothesis 1: Position of Local Realism According to Einstein

There are a number of unknown physical quantities which determine the behavior of entangled particles in advance.

The Uncertainty principle and the assumption of objective randomness are based solely on the subjective ignorance of these variables. Therefore, they cannot be seen as laws of nature. In Einstein's figurative language, "God does not play dice!" Quantum mechanics must therefore be incomplete and needs to be complemented by a theory of hidden variables. Local realism forces us to assume that entangled objects have individual properties that control their behavior. A measurement is therefore always only the reading of a characteristic which has already been predetermined by nature.

## Hypothesis 2: Assumption of Nonlocality According to Bohr

Though spatially separated, entangled particles form an inseparable unit. Before their properties are measured, it is fundamentally uncertain which features will appear in the measurement. What is certain is that they will be correlated. The measurement results are objectively random, i.e. nature in no way determines the properties of the particles prior to their measurement. In particular, quantum objects exhibit nonlocal behavior. Two objects (which may be very far removed from each other) may influence each other instantaneously. According to the local position, they would be able to impact each other at the speed of light at best. This is a fundamental statement of the so-called Copenhagen interpretation of quantum mechanics, which still has a number of followers among physicists.

## Bell's Inequality

We won't go into the complexities of Bell's mathematics here, but rather delineate his inequality in general terms. For deeper understanding, the interested reader is referred to further reading, for example (Zeilinger 2005c). Besides, numerous derivatives exist, as for example the more universal and more easily verifiable CHSH inequality or the "educationally valuable" Wigner inequality (Sect. 2.6.3). The verification of the original Bell inequality can be done by correlating measurement results (such as in the polarization of entangled photons or in the spin of entangled electrons). For this purpose, an extremely large number of entangled particle pairs is measured with respect to their respective correlated properties (spin or polarization direction). This creates the prerequisite for the subsequent statistical investigation, which according to the law of large numbers becomes all the more representative as the number of individual measurements grows. From this, corresponding relative frequencies or probabilities can be calculated, which are then inserted into the inequality. In simple terms, the Bell inequality compares between probabilities that occur with hypothesis 1 and those that can be expected with hypothesis 2. The decisive criterion here is that hypothesis 1 always has to satisfy the inequality. If this is true, local realism would be proven, so to speak, and Einstein would be right. If, however, it is violated (at least in certain cases), hypothesis 1 must be rejected in favor of hypothesis 2.

**Bell Experiments**

Interestingly, John Bell formulated the inequality that bears his name to some degree because he originally wished to support Einstein's position. This effort was all the more important in view of the fact that the EPR problem in the early 1960s no longer aroused any interest and was regarded as a "philosophical skirmish of yesteryear". This assessment has changed fundamentally today. The EPR paper is currently the most cited article in quantum research. John Bell therefore asked one of the first experimental physicists who wanted to check the Bell inequality, the Frenchman Alain Aspect, if he already had a permanent position at his university. Only when Aspect answered in the affirmative would Bell accept Aspect's proposal. In 1982 Aspect succeeded (after earlier measurements by colleagues) with the development of a scientifically even more significant verification of Bell's inequality (doi:10.1103/ PhysRevLett.49.1804). An important point in this and subsequent experiments was to close any possible "loopholes". For example, it is conceivable that the measuring parameters of one measuring device are known at the location of the other. It must therefore be ascertained that the two measurements are separated from each other in terms of time and space (to use technical jargon). This ensures that the direction chosen for one measurement cannot influence the selection of the other, even with signals travelling at the speed of light. Finally, the detection loophole is triggered by measurement errors that always occur in practice. Thanks to computers and modern experimental physics, it is now possible to close one loophole after the other.

**Experimental Results**

Once the results were verified to such an extent that they could be subjected to serious review, they triggered a jolt in the professional world of quantum physics research. It seemed incredible, but it was true. Bell's inequality was significantly violated in all relevant measurement series, thus confirming hypothesis 2 in the sense of the Copenhagen interpretation. And not just in Aspect's pioneering short-distance experiments, but also in the highly media-effective experiments by Anton Zeilinger and Rupert Ursin. Quantum channels were, for example, established in free space across the city of Vienna and even in a sewer underneath the Danube, as well as between the Canary Islands of La Palma and Tenerife. In this way, the violation of the Bell's inequality was already demonstrated over a distance of 144 km. The current distance record (1203 km) in terms of Einstein's "quantum spook" is held by the QUESS experiment, which was described in detail earlier. The "Big Bell Test" (see below) holds the record in terms of high numbers.

An important demand, as mentioned above, concerns the closing of "loopholes". The spatial separation (locality loophole) was already confirmed in experiments by Aspect and Weihs (1998), where an extremely fast manipulation mechanism makes it impossible to transmit information at the speed of light (arxiv:quant-ph/9810080). The detection loophole caused by too low a count rate was closed experimentally in 2001 by M. A. Rowe. The extensive research into loopholes culminated in 2015, when at three different research institutions in three different countries, the TU Delft (the Netherlands), the Austrian Academy of Sciences (Austria) and the National Institute of Standards and Technology NIST (USA), both the locality loophole and the detection loophole were closed in one experiment. The "Big Bell Test" also suggests the exclusion of another loophole. In typical quantum experiments like this, random number generators switch frequently between different measurement arrangements. In theory, the behavior of these generators might be determined by unknown parameters, which means that the setting of the measurement would not be completely free and independent. Therefore, the random decision of more than 100,000 people was used to generate more than 90 million random bits, which were used in 13 different experiments at 12 institutes worldwide to set up the measuring instruments. To be exact, it must be noted that the extreme case of a super-determined universe (in which everything is stringently predetermined and there is absolutely no free will) can never be excluded scientifically even with such tests.

## Conclusions and Significance

The experimental findings on Bell's inequality strongly suggest that hypothesis 1 has to be rejected in favor of hypothesis 2. Everything indicates that Einstein's postulate of quantum theory being incomplete cannot be true. Although Einstein had correctly understood that the Copenhagen interpretation was certainly not compatible with the local realism of classical physics, he was wrong about his incompleteness supposition. Accordingly, a theory of hidden variables capable of reproducing all measured correlations cannot be maintained. This is the essence of the Bell theorem, which most physicists today consider to be proven. Consequently, local realism is equally considered to be invalid. Many physicists believe that at least one of the two principles has to be abandoned, locality or realism. As explained above, "realism" implies the assumption that the measuring instruments only deliver values that have already been determined in advance. "Local" is associated first and foremost with the assumption that the measurement on one particle can at best influence the state of the other particle at the speed of light. Fair enough. The experiment—the highest judge in physics—has spoken. But please savor what all this might mean this in epistemological terms. Schrödinger already stated that entanglement was "… so crazy that it probably compels us to take leave from our beloved, everyday idea of this world". And according to Anton Zeilinger "there is something wrong with our view of the world. Either our ideas of time and space are skewed. Two separated places or points in time may not be separated at all. Or our idea of reality is wrong." (https://www.youtube.com/watch?v=P-f92k-sfKdk) In any case, in the sense of the Copenhagen interpretation, reality manifests itself only through observation (the measuring process). John Bell, of course, was a "realist". He would not believe that his own inequality apparently turned realism upside down. "It's a mystery!" he kept saying. He did, however, give some leeway to the idea that the entire universe might be nonlocal.

Notwithstanding questions of interpretation and philosophical gambits, the simple but far-reaching conclusion (which will become relevant for quantum communication) follows. Quantum mechanics "really" exists, and it is definitely not a theory that can be explained with the help of classical physics. It clearly is, therefore, a nonclassical theory. This demarcation also includes the theory of relativity (which, because of its causal structure, is also rated among the classical theories). Its causal character, however, is what has to be definitely abandoned in quantum physics, as proven by the Bell theorem. Again, it was affirmed that the strange indeterminacy of measurement values does not represent personal ignorance of their true value, but rather that the objects in themselves ("a priori") are indefinite. The wave function only determines the probability of measurement values, but not which concrete measurement result occurs in each individual case. However, it is precisely this position, which is completely contrary to the human mind, that forms the basis of hypothesis 2, which has been confirmed in experiment many times. The fact that entangled objects influence each other instantaneously was also substantiated. Depending on our perspective, we may, with Einstein, deride this "quantum spook". Or—in agreement with scientific positions—we can view it as a basic feature of nature, which per se has to be accepted. The very quality of nonlocality has no classical counterpart and is most evident in the phenomenon of entanglement. Ultimately, however, the philosophical question remains whether explainability (i.e. knowledge) is a fundamental human need that apparently will never be satisfied in the face of modern physics. If it were, what, then, would reality be? It might be, however, that knowledge, i.e. information, represents reality. In any case, there are numerous indications in physics that the concepts of reality and information must not be separated. Irrespective of this, Bell's inequality and its derivations prove to be extremely useful for the future of the quantum internet. This is because it represents a statistical criterion as to whether a quantum channel is intact (e.g. maximally entangled) or whether a technical defect or manipulation has occurred. In this way, any unauthorized interception of quantum cryptographically encoded data can be detected directly. Likewise, objective randomness is confirmed as the most elementary quantum mechanical event. Randomness cannot be reduced further, and this is why the random numbers generated by QKD systems are the best possible that could ever be generated. Besides the fundamental importance for basic research, the technological applicability of entangled states is rapidly gaining importance.

## 1.9    Quantum Information

Before we take a closer look at the term quantum information, let us first clarify what classical information is. A notable example of this goes back to the statistician John Tukey. Placing your order in a coffeehouse, you get a choice of the following variations: hot or cold, large or small, with or without caffeine. With a total of 8 possibilities, the choice seems manageable. The waiter now wants to find out exactly which combination to serve, so he asks, "Do you want your coffee hot?" You answer yes or no. "Would you like a large coffee?" Again, you say yes or no. "Decaf? yes or no?" So, you arrive at an answer after three questions were asked with three yes/no answers.

The information value of this order with its $2^3 = 8$ alternative possibilities is therefore 3 bits of information, which can be represented as a three-digit binary number. This simple idea, namely that information can be symbolized as binary numbers on the basis of yes/no statements, forms the foundation of today's digital technology, and of current IT in particular. The elementary purity of the classical information is represented by the bit (binary digit, unit character "bit"). It contains a choice between two statements (yes or no, true or false), represented as the binary numbers 1 or 0. For $2^N$ questions, the information value therefore is $N$ bits of information. In the computing world it is a known fact that groups of 8 bits, which are then called a byte, are convenient to work with. 1 byte therefore corresponds to 8 questions, which leads to $2^8 = 256$ possible answers or bit sequences. This means that 8 bits of information are required to learn one of the 256 possibilities.

In this way, it becomes possible to "digitize" any kind of information. All one has to do is to ask a sufficient number of questions. One then receives a corresponding number of yes/no statements (bits). These questions may concern anything from the color values of the pixels in a graphics file to the sound pressure values of a music stream during sampling. Typically, a sensor measures a physical quantity (a CCD chip for example gauges the image brightness). The resulting information is transmitted into either a digital signal or an analog signal which is then transformed into a digital value by an analog-to-digital converter. In particular, letters and numerical values can be represented in the form of bit sequences, which can also be used for calculations according to the binary calculus. This dual system (binary system) can be traced back to the "last universal genius" Gottfried W. Leibniz (among others). Today, it forms the basis for data processing in computers based on logic operations (gates). The internal coding of the system, however, depends on the type of information and its subsequent use. File formats play an important role in standardization. Finally, binary values can be stored in working memories, database systems or file systems.

Overall, digital technologies offer numerous advantages, not least because one needs to distinguish between no more than two signal states (0 or 1). These can be realized physically, for example by using a lower or a higher voltage value. Another economic advantage is that the accuracy of the components is rather tolerant, which reduces production costs. This is also helpful for the conventional internet. In terms of physics, the internet it is a complex network system where information processing systems such as computers and mobile devices exchange classical information with each other. In addition to radio-based systems, larger network structures are interconnected across continents, mainly using fiber-optic cables. There are many reasons to choose fiber-optic techniques, first and foremost their enormous transmission capacities. This advantage is primarily due to the fact that light has a very high oscillation frequency. Standard glass fibers typically used in telecommunications function in the infrared range with about $10^{14}$ Hertz. This corresponds to around 100 trillion oscillations per second. It should be noted that fiber-optic systems form an optically dense medium. The speed of light in telecom glass fibers is about 1/3 lower than in a vacuum. This is technically irrelevant, since even bit rates in the range of many TBits are still possible. This is true for each single fiber of course. Several glass fibers together are able to achieve record rates of 1 Pbit/s and more. In graphic terms, the bits are physically represented by very fast on/off light pulses. This "high-frequency technology" is the price we have to pay for digital technology, because it naturally requires very high bit rates per time unit. For this reason, the high "carrier frequency" of light—an ideal physical resource—is put to use. The intensity of light in the glass fibers decreases gradually, however. This means that every 100 km at the latest, it has to be measured, amplified and relayed onwards. Additional technical effort becomes necessary. Despite the technologically complex infrastructure, the basic idea of classical information processing is very simple. The technology has become so ubiquitous that it has led to today's digital revolution.

Because quantum mechanics is so fundamentally different, it has the potential to significantly expand the scope of classical information processing. One distinguishing characteristic lies in the principle of superposition, which has been mentioned before. In the two experiments in Sect. 1.5 we have seen that when a photon passes through a polarizing beam splitter (PBS), it is in a superposition of two possible polarization directions, horizontal or vertical. This can be measured as "0" or "1", which corresponds to a classical bit of information. The essential point, however, is that prior to the measurement, the photon is in a superposition state of horizontal and vertical polarization, i.e. 0 and 1 simultaneously. This forms the basis of a new kind of information—quantum information! In analogy to classical information, whose basic unit is the bit, the elementary purity of quantum information is defined as a quantum bit (qubit). This is the simplest quantum mechanical two-state system, which can assume only two values (eigenvalues) during its measurement. These can be written as 0 or 1. A central question that arises immediately obviously is how much information a qubit can "store" and transmit. This has not been ultimately resolved and remains the subject of current research. There is, however, reason to presume that a qubit contains an infinite (!) amount of classical information. To illustrate this, let's again take a look at a previous example (experiment 2 in Sect. 1.5). The polarization of the light that hits the PBS can be in general be linear, right circular, left circular or elliptical. This means that the electric field strength vector of a light wave can remain constant in a plane, or it can describe circles or ellipses at right angles to the direction of propagation. It can be shown that the field strength vector is always composed of a superposition of horizontally and vertically polarized light (i.e. 0 and 1). It thus represents a superposition of these two basic states. Theoretically, there is an infinite number of possible directions in which light can be polarized. It would therefore also require an infinite amount of classical information to describe all these possibilities. In order to indicate the state of a photon, a classical bit (choice between 0 or 1) is certainly not sufficient. Rather, both parts must be written in a kind of superposition (linear combination). This is the essential characteristic of the quantum bit.

The fantastic thing about quantum theory is that the principle of superposition can be extended to any linear combination of the basic states. We are able to bring as many qubits as we like into superposition with each other. Two qubits in superposition already result in 4 basic states (00, 11, 01, 10), three qubits result in 8 basic states, 4 qubits in 16 states and so on. Because the basic states increase rapidly (exponentially) as the qubit numbers rise, such an arrangement is able to store a significantly larger amount of information than any known supercomputer (depending on which estimate is applied but starting with about 50 qubits). This is also the revolutionary fundamental idea of the quantum computer. If, for example, its functionality is based on the principle of the so-called one-way quantum computer (Sect. 2.5.4), it squeezes all possible solutions to a problem into a very complex quantum state, so to speak—and it does that simultaneously. Then, an attempt is made to read out the solution contained in the state by a skillful series of measurements. The functionality of the quantum computer is supported to a great extent by the phenomenon of entanglement. Einstein in mocking skepticism coined the term "spooky action at a distance" for said entanglement. It enables not only the development of new error correction and redundancy methods (which differ significantly from their classical counterparts), but also the generation of completely new quantum states which cannot be composed of individual sub-states. In addition to the one-way quantum computer, this is particularly important for those concepts which—similar to classical computers—are based on circuit-like models (using quantum gates). Classical computers process information by manipulating bits via logic circuits (gates). A NOT gate for example generates the sequence 10110… from the bit sequence 01001…, essentially inverting the binary sequence in exactly one manipulation. A quantum computer, on the other hand, is capable of using $N$ crossed qubits already for $2^N$ manipulations per gate operation. For $N = 2, 3, 4, 5…$ this corresponds to 4, 8, 16, 32… manipulations per operation. The "exponential effect" becomes evident again, which can lead to a considerable speed-up of the quantum computer compared to a traditional computer. It is not difficult to imagine that a scalable quantum computer (expandable by an arbitrary number of qubits) theoretically promises immeasurable computing power.

## Entropy, Information and Quantum Computers

The possibilities of quantum information can therefore lead to considerable increases in computing speed. In modern physics, however, information is an even deeper concept. This is particularly true for quantum theory.

In nature, processes that cannot be reversed and that run in one direction only occur quite frequently. Let's take a cup of hot coffee as our example. The coffee cools down gradually until a temperature equilibrium with its surroundings is achieved. The cup gets warmer while the coffee gets colder. The cup might also fall to the ground, leaving a pile of broken china. What both these processes have in common is that you will never observe the reverse process. The coffee will never become hotter by itself, nor will the shards jump back together, joining to form the original cup. Such processes in physics are referred to as irreversible. Irreversibility is described by a quantity called entropy. It was introduced by Rudolf Clausius around 1865, who developed it because the principle of the conservation of energy by itself is not sufficient to decide whether a process is reversible or irreversible. Like all physical values, entropy can be assigned a numerical value and a unit (joules per kelvin). It is, however, not the exact value of entropy that is of practical relevance, but only its relative change. In irreversible processes, this change is always positive, meaning it increases more and more until a final state of equilibrium is reached. In the nineteenth century, this idea was so important because it led to the realization that heat can never be completely transformed into mechanical work in the construction of heat engines. Such a machine will never function with an efficiency of 100%. A real-world diesel engine in a car can at best achieve an efficiency of a little over 50%. And that's even using the most modern sensors and electronics.

Still, at the time, it was not really clear what entropy in its truest sense is, exactly. Entropy is also commonly referred to as a measure for the order or disorder of a system, which does not really do justice to this term. While Clausius still somewhat cryptically spoke of the "transformation value", the Austrian physicist Ludwig Boltzmann wanted to get to the bottom of things. What Boltzmann did was to investigate the microscopic aspects of entropy. Although Boltzmann used terms such as ensemble, microstates or macrostates, the result of his investigations is evident today. Entropy has something to do with probability and information.

Let us consider an often-discussed example. Imagine a container which is divided by a central partition wall. The space on the left is filled with a gas. The space on the right contains a vacuum. What will happen if the partition is removed? What will happen, of course, is that the gas will immediately disperse, filling the entire container. The gas atoms or molecules will no longer remain on the left side. Now, on average, 50% of the particles will be on the right side. We now want to know whether a certain particle is on the right side or on the left side. We might ask: "Particle, are you on the left side?" The answer would be yes or no. As outlined above, this corresponds to an information value of 1 bit. We could now ask another particle. Again, we receive an information value of 1 bit, and so on and so forth. With $N$ atoms or molecules we would have to ask $N$ questions (in practice, such $N$-values are unimaginably high). Entropy is simply defined in such a way that it relates to the number of questions. It is therefore qualitatively equal to the number of bits. If we were to ask for the initial state in which all the particles in the container are on the left, entropy would logically be 0 (because we don't have to ask any questions now). But if we ask for the final state, the number of bits would be very large (because now a large number of particles can also be on the right side). Entropy has therefore increased sharply, i.e. changed towards a high positive value. At the same time, however, entropy also describes the tendency of the gas to spread. This is of course an irreversible process, because without external influence, the gas won't automatically converge on the left side of the container. It is intuitively clear that the probability of the gas assuming a state of high entropy must be much greater than the initial state. At Boltzmann's time, of course, the term "bit" did not exist yet. This is why Boltzmann investigated the different states individual particles can assume themselves and associated the resulting number directly with entropy. If these "microstates" (in contrast to a "macrostate" characterized by a few quantities such as volume or temperature) are present in great numbers, the entropy is therefore also very large. Otherwise it is very low. Accordingly, entropy $S$ is both a measure for the information in a state and for its probability $W$. The famous formula which describes this connection is engraved in the tombstone of Ludwig Boltzmann at the Vienna Central Cemetery: $S = k \cdot \log W$. $k$ here refers to the so-called Boltzmann constant. The logarithm (log) contained in it stems from the fact that $N$ bits of information correspond to $2^N$ questions. If we want to derive the exponent $N$, we need the inverse function, i.e. the logarithm.

Now, what does entropy (i.e. information) have to do with quantum physics? Remember that quantum theory began with Max Planck's quantum hypothesis around 1900. Planck found himself virtually forced to formulate it, and he did so with great reluctance. He spoke of an "act of despair". But what had driven Planck into such despair? A major reason for it was that initially, Planck's views were in strong opposition to statistical physics and accordingly to Boltzmann's groundbreaking findings. Only when Planck gave up this position and applied statistical methods in the style of Boltzmann was he finally able to successfully derive the law of radiation named after him. To achieve this, however, Planck had to learn how to "count", i.e. how to divide nature into discrete states, which ultimately led to the assumption of discrete energy packets. And precisely this is the defining element of quantum theory, which stands in stark contrast to classical physics (where arbitrary, continuous energy values are assumed).

Another well-known example is Einstein's photon hypothesis. It is interesting to note that his main inspiration for this hypothesis was his comparison of the entropy of gas (in modern language, bits) with that of light. He noticed an astonishing similarity, which ultimately led him to postulate particle-like photons. Just these few historical examples suggest a close relationship between information and quantum physics. For the prominent US physicist and computer scientist Seth Lloyd from MIT (Massachusetts Institute of Technology), the conclusion is that "all physical objects can be coded into a finite set of bits, which are determined by the laws of nature" (https://www.youtube.com/watch?v=XirbbUxOxiU). If this is actually true, it results in a very significant possibility for the potential of quantum computers. The first quantum simulations are already available today to simulate the behavior of complex atomic and molecular structures. The reason for these endeavors lies above all in the fact that for classical computers, such tasks are often very difficult or impossible to achieve. Because quantum computers work on the basis of quantum information, it is conceivable, given a finite set of qubits, that even structures with much higher complexity could be simulated. To give one example, even a 100-qubit computer could support the development of radically new substances and materials. In the broadest sense we might even speak of "programmable matter". This assessment of the actual technological significance quantum computing represents is in almost perfect agreement with the tenets formulated by Nobel laureate Richard Feynman. Many people, first and foremost US developers and researchers, are inspired by this.

And particularly in the USA, renowned computer and software giants are working with great dedication on the development of the quantum computer. In any case, the interaction of information and quantum physics implies undreamt-of technological possibilities and consequences for humankind.

# 2

# The Quantum Internet

I predict a century of new quantum technologies that will change both science and business. We are only just beginning to understand the possibilities that have started to emerge.

Rainer Blatt

## 2.1 Technological Principles

The term quantum network (casually labeled quantum internet) refers to the networking of quantum information media via quantum channels. Quanta are generally understood to be physical objects that change their state into systems with discrete values of a physical quantity. The term "quantum" is often used for very small units of quantity, for example the smallest amount of light (light quanta = photons) or the smallest amount of energy (quanta of energy). Typically the term is associated with the characteristics of particles, but that is only one aspect of the word's meaning. Somewhat inaccurately, atoms are

© Springer Nature Switzerland AG 2020
G. Fürnkranz, *The Quantum Internet*,
https://doi.org/10.1007/978-3-030-42664-4_2

occasionally informally referred to as "quanta". This designation, however, is not an actual physics term. In information theory, quanta are understood as qubits (quantum
bits) and are then also referred to as quantum information.
They correspond to the classical bits on earlier computers.
Qubits define both the smallest possible amount of memory and a measure for the "transmission" of quantum information. At present, information technology makes very
little use of quantum effects. The future step from bit to
qubit, on the other hand, opens up completely new perspectives. One of these possibilities, quantum cryptography, enables the generation of completely secure quantum
keys for subsequent cryptographic information transfer
using conventional methods via the conventional internet.
Another possibility that is as fascinating as it seems futuristic is the systematic networking of technically feasible
quantum computers.

### Internet Versus Quantum Internet

Today's internet is a complex network of computers where
data is transmitted in the form of classical information
units (bits). In principle, these bits can always be subjected to an eavesdropping attack, or hacked. Data security depends on current computer performance and the
trustworthiness of individuals. In a few years' time, we will
reach a limit for the design and accordingly the performance of existing computer chips, which has its roots in the
laws of physics.

The quantum internet is basically a system of quantum
nodes where a continuous quantum channel is created
until the end nodes are reached. At an advanced stage, it
is a network system of quantum computers that exchange
quantum bits (qubits) via teleportation. The quantum internet is tantamount to a paradigm shift for several reasons.
It is possible to synchronize qubits at superluminal velocity. Also, quantum information is inherently tap-proof and
resistant to hacking attempts due to the laws of physics.
Also, qubits are able to store and transmit much larger

amounts of information than classic bits. Finally, quantum computers have the potential to calculate solutions for problems that even supercomputers cannot solve.

## Quantum Download Via Teleportation

The physics of the quantum internet is fundamentally different from that of the conventional internet due to its novel properties: In the classical internet, mainly fiber-optics technology is used. With its help, information is coded in modulated electromagnetic waves (infrared radiation). These wave trains correspond to periodically varying intensities of a large collective of photons. The information is written in the sheer number of light quanta whose relative change is measured per time unit. Quantum communication, on the other hand, uses the "internal" properties of individual photons themselves. This makes it possible to increase the efficiency of the existing fiber-optic network considerably. Of particular importance in this context is the phenomenon of entanglement. If one changes the quantum state of a system at location A, the state of the entangled system at location B also changes instantaneously (and thus with superluminal velocity). Although the exact characteristic of the state is not determined before the measurement, its relationship to the value measured at A is. An essential characteristic of a quantum internet is that the transmission of information cannot take place via classical repeaters (signal measurement, amplification and transmission). Rather, the quantum information in its entirety has to be transferred from transmitter to receiver. This process is referred to as quantum teleportation. For that, entanglement is a necessary resource. However, anyone who thinks of corporeal "beaming" à la "Star Trek" is far off the mark. It's not matter or even an electric field that is transmitted. Rather, the subject of the transmission is pure information. Quantum teleportation in particular refers to the immediate transfer of state changes of entangled quantum systems. It is therefore necessary to set up a classical channel in addition to the quantum channel.

That quantum teleportation occurs faster than the speed of light seems to contradict Einstein's theory of relativity, which expressly prohibits any instantaneous transmission of usable information. What is decisive, however, is the adjective "usable", i.e. whether the information really is accessible to us. While it is true that the pure quantum transmission is faster than the speed of light, with all the advantages that result from that, it is not possible to interpret these data without a classical channel (for example, a conventional internet connection). The fastest velocity with which any transfer of information in such a channel can take place is the speed of light in a vacuum. And so, consistency with Einstein's theory of relativity is restored. Accordingly, classical channels are an indispensable requirement for quantum IT. It follows that the classical internet can never be completely replaced by quantum networks. It will continue to be essential in communication technology, with quantum technologies providing powerful augmentations. Consequently, the jobs of the many competent and deserving IT specialists working in classical communication industries today are not at risk. In any case, the quantum internet won't destroy jobs. On the contrary, completely new professions will emerge from it.

**Inherent Security**
Closely related to this is another distinctive feature of quantum mechanics, which forms a key element of the quantum

internet. According to the "no-cloning theorem", it is impossible to copy quantum states perfectly. The emphasis is on "perfect". It's therefore also impossible to duplicate original qubits and make an arbitrary number of copies (see also Sect. 3.6). This is the special secret of the inherent data security due to the laws of physics. For an eavesdropping attack to be successful, the information must exist for the eavesdropper independently of the sender. This implies that it must be at least doubled. It is, however, impossible to duplicate qubits and consequently, as they cannot be intercepted. Duplication of qubits would be a definite violation of Einstein's theory of relativity. (Since qubits are transferred instantaneously via teleportation, this would be equivalent to information transfer at superluminal velocity. The theory of relativity, which has been proven millions of times over, categorically excludes this.) As will be shown in Sect. 3.6, the no-cloning theorem is also consistent with the special theory of relativity. Not least because of the no-cloning theorem, quantum systems react with very high sensitivity to external influences and break down easily. This is why typical hacker attacks do not stand a chance. Cracking firewalls, smuggling in trojans, scattering viruses etc.—none of these are possible with a quantum computer!

## Quantum Repeaters

And so, the no-cloning theorem poses a genuine challenge in the development of quantum networks. It has far-reaching consequences for quantum informatics. For example, it is not possible to use classical error correction systems or, even more importantly, conventional repeaters. Instead, it becomes necessary to develop special quantum repeater systems. A quantum network basically requires two types of qubits: on the one hand "dormant", so-called stationary qubits (e.g. for quantum memory systems) and on the other hand mobile qubits that transmit quantum information to other nodes. In order to effectively perform teleportation, all qubits—both stationary and mobile—have to be in superposition. They are, so to speak, compressed into a single quantum object that is connected to itself as if by magic. The elementary problem here is the production of entangled systems over longer distances. This cannot be achieved without a quantum internet. For its realization, the existing fiber-optic network has to be modified drastically, which is done by equipping it with countless quantum repeater nodes.) The technical term for this is "entanglement swapping". This method makes it possible to extend the entanglement of many smaller subsystems over long distances. Because of the sensitivity of the quantum states (no-cloning theorem), however, the technical realization is extremely difficult, although the fundamental mechanism has already been proven under laboratory conditions.

**Scalable Network Expansion**

The developers' spec sheets contain numerous requirements for evolutionary network expansion (scaling). Some details: the projects are in principle to aim at uniform standards and protocols which are sustainable internationally. In particular, this involves the development of suitable repeater systems.

The essential building blocks for this have to be designed on the basis of relevant repeater architectures and fulfill the following requirements: storage of quantum states with high fidelity (correspondence between input and output states), purification of entangled states (entanglement distillation, Sect. 2.8), implementation of quantum logic gates, conversion between mobile and stationary qubits; also, suitable communication protocols and addressable quantum registers for the transfer of quantum information. For the realization of these components, various approaches are applied, all of which are currently being investigated scientifically. The most promising physical resources to date, besides photons, ions and ultracold atomic gases, include semiconductor structures, superconductors, quantum dots and hybrid systems. Special attention is also given to the use of new materials such as graphene, but also to color centers in diamond-like lattice structures with promising coherence properties. Further details will be described in the following. In general, the development of quantum networks is closely linked to the realization of quantum computers, the latter being a much greater technological challenge.

## 2.2 Network Topology

The world of the internet, as fantastic as it is even today, is based on sophisticated network systems that transmit digital information between different nodes worldwide. In a hierarchical arrangement, provider, company and research networks are connected to a global web via backbones, mainly using fiber-optics elements. The development of technologies that are suitable for a quantum internet poses enormous challenges. One promising concept is a network system consisting of quantum memories. Between those stationary network nodes, quantum information is reversibly exchanged over long distances via mobile qubits (mostly photons).

The basic infrastructure of a quantum network corresponds well to its classical counterpart (Fig. 2.1). The actual operations are performed at the so-called end nodes. One strives for a system where these are connected to each other via entanglement. In the simplest case, an end node consists of a single qubit, which develops increasingly into a

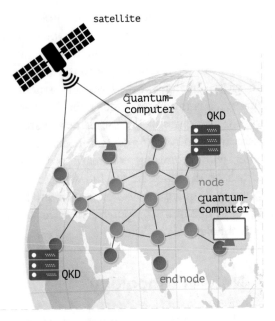

**Fig. 2.1** Topology of a quantum internet. The network is characterized by a system of quantum nodes that produces a continuous quantum channel end-to-end

powerful quantum processor as the number of qubits rises. For more straightforward applications such as quantum key distribution (QKD), the end nodes are equipped with comparatively simple devices. Some protocols, however, require much more complicated nodes. These systems enable higher processor performance and can also be used as quantum memories. Quantum logic operations can also be performed on them. In order to transport quantum information from one node to another, special communication lines, the so-called quantum channels, are necessary. Naturally, there is a tendency here to aim at compatibility with existing fiber-optic connections wherever possible, the quality of which is usually sufficient for QKD operations. In order to ensure efficient communication, routers and switches are required—just as is the case with the classical internet—to switch the qubits to the desired end nodes. However, switches have to be able to ensure quantum coherence for a relevant period of time (which is generally very short), making their technical implementation much more difficult than with today's standard devices.

Open space networks operate in similar ways as fiber-optic networks do. However, they use a direct sight line (free beam path) between transmitter and receiver, typically a direct laser connection. As recently demonstrated by the QUESS experiment, quantum satellites communicating via quantum channels in space can also be used. First and foremost, such satellites offer the possibility to create direct entanglement, without the need for quantum repeaters, over larger distances. In the future, they may also play an important role in linking smaller, ground-based networks over long distances. With a globally distributed satellite system and corresponding logistics, worldwide networking is also conceivable. In purely theoretical terms, quantum satellites can act as quantum repeaters for short periods of time.

Such a network could be designed either as a quantum network for computing or as one for communication. In the first case, different quantum processors would be connected to form a quantum computer cluster. This is referred to as networked or distributed quantum computing. In this case, less powerful quantum processors are linked. The result is a much more powerful quantum computer. This is analogous to classical computers that are interconnected to form a cluster. Networked quantum computing is often regarded as a possible way for the realization of a scalable quantum computer, because an increasing number of interlinked quantum processors will increase computing power in theory. In earlier approaches to networked quantum computing, the individual processors were often separated by tiny distances only.

A quantum network for communication, on the other hand, offers, among other things, the possibility of transmitting qubits over long distances from one quantum processor to another (long-distance quantum communication). In this way—analogous to the classical internet—smaller networks can be interconnected to form a larger one, which ultimately makes a global quantum internet conceivable. This would enable a wide variety of applications, whereby the performance, in addition to the processor capability of the nodes, is decisively determined by the extent to which entanglement can be generated and maintained.

## 2.3 Quantum Interfaces

Today's internet already sends an enormous amount of data around the globe, mainly via fiber-optic cables. The quantum networks of the future will be much more powerful because they exchange quantum bits that can carry and transmit much more information. However, an essential requirement for this are components with which the quantum information can be reversibly transferred from a quantum memory to mobile qubits. The term "interface" generally refers to central transfer points that transmit data between computers and external devices. A quantum interface is a device that connects stationary qubits with mobile qubits in order to create a quantum channel between distant nodes. Although in reality the processes are very complex, this can be described (in vastly simplified terms and using descriptive language) also with familiar phrases: At node 1, quantum information is stored in a stationary quantum memory (q-memory). This is then "read out" and transferred to a mobile qubit, which moves at the speed of light to node 2 and writes the quantum information into a q-memory present there. The two nodes are brought into superposition in this way. The quantum info is not copied, but the generated entanglement is a single common state, which means that the no-cloning theorem is not violated. This process can be repeated in any direction (reversibility). However, even though this sounds simple and is a standard procedure in typical IT operations, is much more difficult to realize for a quantum network. For this reason, intensive research is being conducted worldwide into how efficient quantum interfaces can be implemented. It is also essential how precise and controlled the operations can be carried out without destroying the extremely sensitive quantum states that arise.

### The Cardinal Problem of Decoherence

As will be shown later (Sects. 2.5, 3.1, 3.3), quantum objects are subject to the superposition principle, which forms the basis of entanglement. If the mountains and valleys of two waves have a fixed phase relationship to each other or change over time according to the laws of physics (see also Sect. 3.2), they are referred to as coherent. The same is true for the wave functions that describe quantum states. However, since quantum particles inevitably interact with their environment, the phase relationships (relative distances of the wave crests and valleys from each other) get out of sync, and coherence is lost. Due to this decoherence, the quantum world loses its typical properties and enters the field of classical physics. Technologically, this means that extreme caution must be exercised with regard to external influences, i.e. systems must be developed which guarantee coherence for a sufficient period of time in order to be able to carry out quantum mechanical operations on them. On the other hand, the quantum systems also have to be manipulated, measured and read out. This is the real technological challenge in the development of a quantum network.

## 2.3.1 Nobel Prize for Preliminary Work

The work of David Wineland and Serge Haroche, who were awarded the Nobel Prize in Physics in 2012, formed an important basis for quantum information technology and especially for quantum interfaces. The two scientists developed groundbreaking experimental methods for measuring and manipulating quantum systems.

**Laser Cooling**

Atoms are known to be tiny little objects. In order to be able to control and manipulate them, we have to use special "tricks". A very strong vacuum, for example, will cause the pressure in the environment to drop to practically zero. The few individual atoms still present can be easily trapped with suitable devices. For example, a voltage is applied between specially arranged electrodes. This results in potential wells where the atoms remain trapped just like golf balls in their holes. The particles can remain ensnared there for longer periods of time. In order to achieve particularly strong quantum coherence, it is also necessary to cool the particles very strongly, which further reduces their thermal motion. This is where Wineland's contribution comes into play. Even when they closely approach absolute zero ($-273.15$ °C), the particles still oscillate, though not freely, but only in certain directions, which are determined by the laws of quantum mechanics. Using a special type of radiation, it becomes possible to slow down the particles further, bringing them into their lowest possible energy states. For this purpose, a laser beam is directed at the opposing atom, energizing it into an excited state. During the subsequent emission, the discharged light quantum generates a recoil, which slows the movement of the atom. This means that it loses energy in the direction of the laser beam towards which it has been moving. David Wineland perfected this method and was able to demonstrate that quantum objects can be controlled, manipulated and read with high precision using targeted laser pulses.

## Cavity Resonators

Serge Haroche, for example, came up with the idea of a special optical resonator. The setup consists of two mirrors. Single photons are reflected back and forth between them. However, only such photons can participate in this game of ping-pong where multiples of their half wavelengths fit exactly into the distance between the mirrors. While at ambient temperatures, innumerable types of oscillations are possible in the resonator, only very few remain near the absolute zero point. Under these exotic conditions, specially prepared atoms can be introduced into the resonator, causing a specific interaction between the atom and the electromagnetic field of light.

## Cavity QED

Quantum physics is basically structured into quantum mechanics and quantum field theories. The most commonly known of these is quantum electrodynamics (QED), which explains electromagnetism as an exchange interaction mediated by photons. Cavity quantum electrodynamics (cavity QED) investigates the interaction of light trapped in a reflective cavity, such as an optical resonator. With the help of such cavities, quantum interfaces or even quantum computers can be constructed. If the light in the cavity is in resonance with an atomic transition, coherent exchange with the field of the cavity occurs. This can result in entanglement between the atomic state and the cavity field. Physicists Wineland and Haroche played a major role in this development. With the help of cavity QED, the principle of a quantum interface and its coupling is shown here in strongly simplified form (Fig. 2.2).

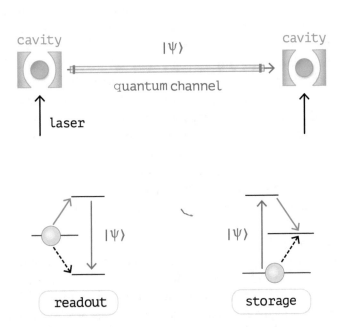

Fig. 2.2 Principle of node superposition via quantum interface

At node 1 in Fig. 2.2, a single atom is trapped in a cavity. It is in a medium energy state. The atom is then excited by a laser, so that the quantum state of the atom briefly assumes its highest energy state. Immediately after that, the atom "falls" to its lowest level, resulting in a quantum state that is transferred to a photon (mobile qubit) and transferred via a fiber line to the identical node 2. There, the atom is in its lowest state. It is excited by the incoming photon, changes briefly to the higher state and then falls back to the medium level. This creates an entangled channel between node 1 and node 2. In descriptive terms, this corresponds to a reversible storage and readout process of quantum information via spatially distant nodes.

## 2.3.2 Implementation (Examples)

To give the reader an impression of the future "quantum IT", a few implementation attempts for quantum devices will be mentioned here. They will at least give you an inkling of both the potential and the challenges this exotic technology brings.

**Trapped Atoms**

Already several years ago, Rainer Blatt and his research group in Innsbruck succeeded in realizing the prototype of an elementary and largely controllable quantum interface. A charged atom, for example a calcium ion, is trapped in a so-called Paul trap and placed between two highly reflective mirrors, i.e. in an optical resonator. A laser excites the ion and entangles it with the photons of the laser. The frequency and amplitude of the laser can be used to influence the degree of entanglement in a targeted manner, making it possible to optimally adjust the number of entangled photons collected in this way. How can one imagine this process, which corresponds to a registered letter of quantum information from a stationary qubit to a mobile qubit, in concrete terms? To illustrate this, here's an example scenario. Let's assume that the electrons orbit around the atomic nucleus according to the Bohr atomic model (of course, this only serves as a working aid for illustration which, strictly speaking, is not correct in terms of physics). The superior calculation power of quantum computers lies precisely in the fact that a superposition of the two states is generated, i.e. 0 and 1 at the same time. This excitation corresponds, so to speak, to a superposition of both electron orbits. The excited state is then entangled with the polarization state of the laser photon. This means that the overall state can no longer be separated into individual sub-states, it can only be described as a whole. When the mobile qubit moves through a fiber line to a second interface, it "carries" the mutual quantum information, which is determined by the entanglement between atom and photon, to this second node. Just recently, researchers in Innsbruck set a record for the transfer of quantum entanglement between matter and light over a distance of 50 km in fiber optic cables.

### First Prototype Network

Gerhard Rempe and his team at the Max Planck Institute in Garching succeeded in realizing an elementary quantum network of 2 nodes as early as 2012. They did this by coupling two cavity resonators. The extraordinary innovation was that the team were able to entangle massive objects, more specifically, atoms. Casually speaking, the Garching setup corresponds to a kind of nanoswitching system consisting of atoms, which acts like a transistor in a microprocessor. Because of the entanglement of the nodes, however, the system is able to bridge a greater distance, just like a single synchronized switch that functions both as data carrier and arithmetic unit. First, the atoms had to be contained in the resonator for a longer period of time. Using finely tuned laser beams, the atoms were excited in order for them to emit light quanta. In this way, it was possible to reversibly store and read out the quantum information inscribed in the photons over a longer period of time. Due to this symmetrical behavior, the system appears to be well suited for networks of many resonators. Similar to Fig. 2.2, node 1 generates entanglement between the atomic state and the polarization state of the emitted light quantum. During the absorption process, this entanglement is transferred to the atom at node 2. In this way, it was possible to demonstrate the entanglement of atoms over a 60 m long glass fiber connection. Even though the lifetime of the entanglement was only about 100 μs, this was a much longer time span than was required for generating the channel. Meanwhile, the researchers also succeeded in transferring quantum states to ultracold atom gases or Bose-Einstein condensates (BEC). Likewise, they were already able to use the resonators for quantum logic gate operations. This is an important step towards networked quantum computing. The term BEC refers to a larger atomic bond that transforms completely into a collective quantum state when its temperature falls below a very low transition temperature, i.e. reaches the "ultimate entanglement". The entire atomic structure is then described by a single wave function and behaves ideally coherent. The behavior of BEC is closely related to that of superconductors, which is why such controlled ultracold atomic gases could be used to simulate unexplained processes in superconductors and thereby investigate them in detail.

## Quantum Chips and Diamonds

Researchers from the renowned American MIT are collaborating with Harvard University scientists on a novel combination of quantum communication and traditional chip technology. One of their main research objectives is the development of a scalable quantum interface. As outlined above, the essential challenge is to capture and manipulate atoms in a controlled way. This can also be done in a "natural" way, for example with the help of diamond-like structures, i.e. an atom trap in a modified carbon lattice. One such component which has already been developed consists of so-called NV-centers[1], which function as quantum memories. Each NV center stores the quantum information in a combination of electron spin and nuclear spin. Various other spin states are required for error correction. A special integrated circuit routes the NV photoemission, on the one hand for detection, on the other hand for the interconnection with mobile qubits for the network. NV centers generally exhibit suitable properties for quantum memories, including long spin coherence times (1 s, which is an eternity in quantum IT). They can also be used for 2-qubit gates or quantum error correction systems. As early as 2015, the entanglement of two NV centers over a distance of 1 km was demonstrated as an important step towards quantum interfaces and for networked quantum processing.

### Hybrid Quantum Nodes

As the previous examples show, different types of implementations are possible.

Quite similar to today's internet, which also combines myriads of different devices in one network, the future quantum network will also connect various quantum devices. This is why many researchers today believe that hybrid network nodes will be of high relevance for future developments, and not just the integration of completely similar interfaces. As some node implementations seem to be more suitable for certain tasks than others, a quantum network would benefit from the ability to access different types of nodes. Ultracold atomic gases, for example, are able to generate qubit-coded photons without difficulty, whereas doped crystals are well suited for the long-term

---

[1]In an NV-center, two neighboring places in a diamond lattice are not occupied by carbon atoms, but rather by a nitrogen atom (N) and an empty space, a vacancy (V).

storage of quantum information. However, different nodes emit and process photons at different wavelengths and bandwidths, which makes the qubit transfer between them much more difficult. A group of researchers headed by Hugues de Riedmatten (ICFO Barcelona) was able to demonstrate an elementary hybrid node connection. Specifically, a laser-cooled cloud of rubidium atoms operated as a stationary qubit. This was then encoded in a single mobile qubit, a photon of 780 nm wavelength (nm refers to nanometer, which corresponds to one billionth of a meter). The transfer was carried out between two adjacent laboratory stations, reducing the wavelength to 606 nm so that it could interact with the receiver which consisted of doped crystal nodes. In the meantime, the photon used in this process was converted to the "IT-standard" of 1552 nm in order to show that the technology is in principle compatible with conventional telecommunications infrastructures. In this elementary demonstration of the interaction between different quantum nodes and the use of their respective advantages, the researchers see an important milestone for the development of a fiber-optics based quantum network.

## 2.4    Possible Applications

The aspects of a quantum internet are as diverse as they are unpredictable. This is why it is not yet possible to present an overview with any claim to completeness. As always at the beginning of a radically new technical development, nobody can predict with any certainty how things will evolve. Far too often, physicists and technicians don't agree. Some relevant basic functionalities can already be identified today however, and several near-term developments can be inferred (see also Sect. 2.9).

## 2.4.1 Protection, Coordination and Processing of Data

At the moment, the most significant and most advanced application of the quantum internet is quantum key distribution (QKD). Particularly in connection with protocols based on entanglement, QKD combined with classical technologies has high future potential. This assessment is supported by the fact that first QKD systems have entered the market. Devices that establish tap-proof point-to-point connections and products intended for use in real network environments are already commercially available. Major science projects for the development of QKD systems have been active for years. Direct entanglement over long distances is not practicable yet, which is why previous efforts have, among other things, focused on demonstrating the infrastructure and the functionality of the devices. Such systems are typically referred to as "level 0" QKD networks on the basis of "trusted repeaters". These systems were initially tested in laboratory studies, and now first prototypes have been implemented in real networks in urban areas. Prime examples for such metropolitan systems are the QKD networks in Japan and China, which will be discussed in more detail later (Sects. 2.4.2 and 2.4.3). According to the manufacturer's information, the first commercially usable trusted repeater network in the USA is under construction, providing a QKD channel between Boston and Washington D.C. over a total cable length of 800 km.

Larger quantum networks however, where several qubits are connected to their end nodes via entanglement outside the laboratory environment, have not yet been implemented. While this is much more difficult, it is also much more exciting. The application possibilities known so far include, for example, coordination of distributed system problems, clock synchronization, position verification and very long baseline interferometry for radio astronomy (to achieve higher resolutions). One example is the synchronization of atomic clocks. Even today, precise time recording is carried out internationally via a global network of atomic clocks synchronized by satellite. Ultra-precise optical atomic clocks exist today that exhibit an error of much less than one second when calculated on the age of the universe (approx. 13.8 billion years). To make use of this enormous precision, however, or to compare the performance of these clocks, a satellite connection is not adequate, because the "noise effects" caused by the link cancel out these advantages. Scientists have been able to demonstrate that optical clocks can be entangled via fiber-optic cables. The ultimate solution would be a global quantum internet of optical atomic clocks, which would result in a profusion of ultra-precise clocks ticking in complete synchroneity all over the world. A quantum network benefits from the fact that, although it cannot communicate faster than light, it can coordinate and synchronize itself at a speed that is well above the speed of light. Especially the latter (which would be completely impossible in a classical internet) makes quantum communication so valuable and interesting. Already today, countless coordination problems occur in classical networks, which will require much faster and more efficient solutions in the future. For this purpose, quantum bits exploit the fact that they are automatically and instantaneously coupled to each other via entanglement. Likewise, the initial state of a quantum processor can be transferred by quantum teleportation to another quantum computer, which uses it as input. This results in much higher data rates.

An important application arising from this technology is distributed or networked quantum computing, where an entire network merges into a single computer. Already demonstrated on a small scale as a mini quantum network, such a system might theoretically be extended to larger distances. This has the advantage that more complex processors, which would be manufactured at different points on the planet using different technologies, could be connected with each other to form a single quantum mainframe. Already now, it is foreseeable that a modular structure would have many advantages also for a future quantum computer. To achieve this, many quantum processors, each of which is only able to store and process a limited number of qubits, are interconnected via quantum channels. According to an approximate estimate, such a "combined system" would be able to solve certain problems of high complexity faster than classical computers, starting already at a number of about 50–60 qubits. Note that each additional qubit increases the computing power exponentially (!). The top performance of future quantum computer networks will be determined by two factors:

1. Which problems can be solved by quantum computers in the first place?
2. How could the (possibly very different) capabilities and concepts of individual computers be combined in such a way that even more complex problems can be solved?

As will be discussed in detail, the known potential of quantum computers today, besides search and logistics algorithms, lies primarily in calculations where the number of arithmetic operations increases exponentially, as well as in the simulation of atomic and molecular structures. The latter in particular could have a lasting effect on the ecological and economic development of humanity, even beyond research. In view of the above, the possible applications of the quantum internet can be seen in four basic functions:

1. tap-proof communication via QKD,
2. synchronization at superluminal velocities and coordination of quantum systems,
3. quantum communication between quantum processors in the sense of networked computing or a modular quantum computer,
4. multi-user access to a quantum cloud.

The last point means access to powerful cloud quantum computers for many people around the world. These central computers, which would of course be owned by relevant companies or institutions, may again be implemented as local quantum networks. As far as safety is concerned, the quantum internet sets completely new standards for each of the aspects mentioned.

## 2.4.2 Tokyo QKD Network

Complementing the usual security technologies, quantum cryptography represents a highly secure alternative. Relevant systems have been increasingly developed since the 1990s, and prototypes have been released from laboratories into real network environments since the 2000s.

In so-called QKD field tests, it was analyzed to what extent this technology is suitable for practical use and how it can be employed in various applications. In 2010, nine organizations from Japan and Europe collaborated on the largest QKD test to date. The primary goal was to demonstrate the suitability of high-end security technology for commercial use. Relevant applications include secure TV conferences and mobile telephony. While in earlier experiments, bit rates did not exceed a few Kbit/s and distances were limited to about 10 km, much higher bit rates over distances of about 100 km were demonstrated in the

Tokyo field test. The exceptionally high generation rate also made real-time encryption possible. Parts of the former NICT test network were adapted to create the Tokyo QKD. It has four main access points which are connected by commercial fiber-optic lines. These are located in the towns of Koganei, Ōtemachi, Hakusan and Hongō on Japan's main island Honshu. The large distances within this so-called metropolitan network posed the challenge of considerable losses in those long fiber-optic lines (about 0.3–0.5 dB/km). The unit decibel (dB) is a logarithmic ratio value for gauges and measurements. In the case of light, this value refers to brightness and therefore light intensity. In the quantum model, this corresponds to the number of photons. Optical attenuation (loss of intensity), environmental influences or "crosstalk" caused by adjacent fiber lines in the same cable cause significant noise effects. This implies that considerable technological and scientific know-how is required to be able to compensate the losses in the best possible way. The members of the consortium were NEC, NTT and Mitsubishi in Japan, and Toshiba Europe UK as well as Switzerland-based ID Quantique in Europe. Further support was provided by the "All Vienna" team consisting of the Austrian Institute of Technology (AIT), the Institute of Quantum Optics and Quantum Information Vienna (IQOQI-Vienna) and the University of Vienna. All organizations used quantum devices of various types each of them had developed separately. In this way, a node mix type based on (mainly) trusted repeaters was created.

**Trusted Repeater**

As mentioned above, the creation of a quantum internet requires direct entanglement between all end nodes. To achieve this, special quantum repeaters have to be developed. This will probably take some time, and so larger network systems based on what is commonly referred to as "trusted repeaters" are currently being implemented as an intermediate step. Such a system can be compared to a relay race, where a quantum-secured handover is done at each point. It is, however, necessary to assume that each transfer node is "trusted" and does not share any information without authorization. Asymmetric or hybrid procedures are used to ensure the highest possible level of security within the trusted nodes (see Sect. 2.6.1).

**Further Details**

Between the end nodes A and B of a secure data link, the trusted repeater R is installed. First, two private keys $k_{AR}$ and $k_{BR}$ are generated. A then sends a key $k_{AB}$, which has been encrypted using $k_{AR}$, to R. R deciphers the key and gets $k_{AB}$. R then encrypts the key $k_{AB}$ again using $k_{RB}$ and sends it on to B. B uses $k_{RB}$ to decipher the key and gets $k_{AB}$. The resulting mutual key $k_{AB}$ can now be used for transmitting data via a conventional IT connection. The system is absolutely safe from attackers outside connection A and B, but not within the transmission channel, because R is able to decipher all keys and therefore hast to be trustworthy.

**Network Architecture**

QKD networks are arranged in a hierarchical structure that consists of three levels. The lowest level is the quantum layer. It is based on special relay stations via trusted nodes. Each link generates the security key in its own way. The protocols used and the formats and sizes of the keys are different from each other as well. In the most cases, various decoy-state-BB84 systems are used. The big exception is the "All Vienna" group. The Vienna system is based on entanglement. Via QKD devices, the keys are transferred to the central level, the key management (KM) layer. Here, a key management agent (KMA) receives the keys via an application interface developed by NEC and NICT to be compatible with the system. The KMA is a classical computer that works as a trusted node. Its job is to identify the key material and to adapt it in size and format to the requirements. It then stores the keys in numerical order so as to synchronize key usage for encryption/decryption. It also stores statistically relevant data such as the quantum bit error rate (QBER) and the key generation rate. It then forwards the resulting information to the key management server (KMS), which is responsible for central network management. The KMS handles and coordinates all connections. All network functions run entirely on the KM layer under the control of the KMS. The server also monitors the lifecycle of a key and notifies secure paths. The authentication procedure is performed according to the so-called Wegman-Carter scheme, which is based on previously generated keys.

Finally, the third level guarantees secure communication due to the specially generated quantum keys for encrypting/decrypting text, audio or video data. The users are inside the trusted nodes. Their data are sent to the KMA and encrypted/decrypted using the so-called one-time pad (OTP) process (Sect. 2.6.1) in a stored key mode. Because there is a limit to the number of relay stations in a mixed type network, the KMS establishes a route plan for the endpoints of the user requests and selects the appropriate route. Autonomous search algorithms are used for this purpose.

**Demonstration**
In October 2010, the successful functionality test was presented to the public. This was done by a video conference between Koganei and Ōtemachi. The live video stream was encoded quantum cryptographically and then transmitted

in stored key mode via OPT. The key rate was 128 Kbit/s. The team succeeded in generating secure keys over a total distance of up to 135 km. On a 90 km long connection line, a hacker attack was then simulated as a special security check. For this, the link was violently attacked by a laser beam. The KMS detected this attack through the immediate increase of the QBER and gave alarm. To this end, KMAs Koganei 1 and Koganei 2 provided keys that had been stored for that purpose. The KMS immediately switched to a backup line, in order to continue with the key generation before the keys ran out. The TV conference then continued without further disturbances and the safety remained intact. In addition, several switch tests for the various relay lines were performed with success (https://arxiv.org/abs/1103.3566).

## 2.4.3   2000 km High-End Backbone

After the Japanese QKD network was successfully presented to the public, ambitious China would not be left behind. China has taken things one step further. With the Beijing-Shanghai project (China-QKD network), a distribution network of around 2000 km was implemented. In the future, this might form the backbone of the national communications network. From Beijing in the north via Jinan and Hefei to the coastal city of Shanghai, this backbone connects four metropolitan quantum networks. Based on similar technology as in Japan, the Chinese network exceeds its Japanese predecessor not only in its greater length but also in the fact that the QUESS experiment demonstrated quantum satellite technology for the first time. This provides the essential option of directly interconnecting distant end nodes by means of quantum satellites and thus raising QKD to a considerably higher level of safety. According to media reports, the network is already being used by administration, members of government, financial players and the military. The Jinan Institute of Technology also states that the network is ready for commercial use. The high-end communication line is also much more secure than conventional telecom connections, which don't possess an inherent physics-based security mechanism. Thus, China hopes that the pilot project will extend beyond the country and successfully spread throughout the world. Since 2014, the world's largest QKD network has been under construction. The first phase of the project was completed in September 2017. A bank transaction from Shanghai to Beijing which was performed during the opening ceremony generated considerable media attention. An additional connection was established in Wuhan, the capital of Hubei Province. This development is to be continued with further urban quantum networks along the Yangtze River. Finally, it is planned to push the expansion even further with an additional several thousands of km. Of course, security is to some degree limited by the fact that the Chinese quantum network, like the Japanese one, works primarily with trusted nodes. More precisely, it consists of a chain of communication lines connected by 32 trusted nodes (as of 2017). As mentioned above, this does not yet result in the best possible protection. Nevertheless, at this project stage, the system is already much safer than classical networks because the theoretically infinite number of interception points is reduced to 32. At any rate, the local researchers are hoping for a global quantum network, which, according to scientist Jian-Wei Pan, will be deployed at the end of the coming decade. However, the researchers' communication on technical details still remains somewhat restrained.

## 2.4.4 The Vienna Multiplex QKD-Web

One of the main motivations for setting up a quantum network is to make QKD available to as many users as possible. Since many implementations are designed for no more than two parties, one of the most important research directions is to develop the most efficient multi-user solutions possible. Austrian researchers under the direction of Rupert Ursin have recently taken a groundbreaking step in this direction. In a proof-of-concept study, the team was able to demonstrate a resource-saving and scalable network architecture that promises considerable advantages in speed over previous designs. Just as sensational is the fact that the new system is a fully interconnected QKD system and thus a "real" quantum network. As will be discussed in detail later, this implies a significant increase in the security level as compared to trusted repeaters. In the test, it was shown that a single passive entanglement source enables high-security QKD between four communication partners via frequency division multiplexing. Specifically, a special laser generates an entangled polarization state. The frequency spectrum of the light is then split into 12 channels via bandpass filters, three frequencies of which are then allocated to each of the four users at the end nodes via an optical fiber. The frequencies are switched by this multiplexing in such a way that one partner always shares an entangled photon pair with the other. In this way, all six possible combinations with four participants can be realized. The design is particularly cost-effective and user-friendly, as the end nodes are equipped with comparatively simple devices for consumers. The researchers confirm that the Vienna architecture can be adapted directly to any other network topology and that it will be linearly scalable. New clients can therefore be added to the system with minor modifications. By using wavelengths in the "telecom standard" range, the network is compatible with the existing infrastructure of the internet, making it one of the most promising implementations for a commercial QKD network (https://arxiv.org/abs/1801.06194).

## 2.4.5 The Quantum Cloud

Already today, cloud computing is an important component of the internet that is utilized by many users, especially companies. Among its services are infrastructures such as application software, computing power or storage space for remote users. However, the central collection and evaluation of user activities (keyword big data) raises major concerns regarding data protection, which will become all the more urgent in view of the future development of the internet. Even solutions such as homomorphic encryption still offer targets for attack. A quantum cloud, on the other hand, is not only able to contribute to the speed-up for certain numerical tasks, it can also support unprecedented levels of security. Concepts such as privacy or integrity will be taken to a new level: After all, a highly secure client-server environment is created where all of the calculations performed by the server remain completely unknown. This functionality does not exist in classical IT. The idea behind this leads to the key concept of blind quantum computing (BQC). An example: one would like to use the advantages that a quantum computer offers for certain applications. This, however, would require very complex and cost-intensive hardware, which would make it too expensive and difficult for most users to manage. It's a different story with a quantum cloud. There, the client simply requests computing time. After receiving the request, the quantum server connects to a quantum processor, which then performs the requested tasks. It is essential that the quantum computer calculates "blindly", i.e. the client only gives instructions which the server executes. For fundamental physics reasons, the user cannot have any information about the process. It is quite conceivable that a future quantum internet will offer cloud access to very expensive and powerful quantum computers, which would typically be owned by major companies and institutions. Thanks to BQC, not only would a considerable level of security be guaranteed, it would also be possible to check whether the calculation is really done by a quantum processor and not a forgery.

Numerous proposals and studies exist today regarding the implementation of a quantum internet. One possibility, for example, is for the client to make qubits that were generated by a cluster available to the server. The server has the resources to perform the required calculations according to the instructions the customer transmits in the conventional manner. Photons, for example, are suitable as mobile qubits because they can be transported through glass fibers. The server now entangles these qubits via special quantum transformations without any information about the kind of entanglement. Then, the server performs measurements according to the principle of the one-way quantum computer (according to instructions received from the customer; Sect. 2.5.4). However, the measurement results by themselves cannot be used directly by the server, because the random element of the qubits introduced by the client remain an unknown to it. When the server now sends the results back to the customer, only the customer is able to interpret them correctly, because nobody else knows the objectively random measured values of the customer's qubits. All in all, the system is based on the principle that the server can never have access to the full information about the quantum states of its client.

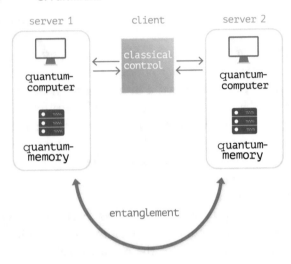

**Fig. 2.3** Blind quantum computing

In recent years, several different BQC protocols have emerged, all of which require clients to have access to special quantum devices (either for preparing or measuring the qubits). A Chinese research team at the University of Science and Technology in Hefei under the direction of Jian-Wei Pan and Chao-Yang Lu was able to show that it is even possible for a purely classical computer to delegate a quantum computer to a quantum server (Fig. 2.3). Using a simple example—the division of the number 15 into its factors 3 and 5—it was demonstrated that a purely classical client is able to communicate with two quantum servers without these being able to know exactly what is being calculated. This becomes possible because either of the two servers only performs parts of the calculation and it is physically impossible for the servers to exchange information with each other in the classical way. At the same time, the result of the quantum computation can be checked for "cheating". With their proof-of-principle demonstration, the researchers hold out the prospect that this method could be upgraded to "real" computer problems and one day implemented in cloud servers. According to Lu, another big advantage lies in the fact that the user does not need any special (possibly expensive) quantum devices. This saves resources and theoretically makes scalable quantum computing available worldwide. A fantastic and at the same time feasible vision: Quantum power hat is being distributed discreetly and without possibilities of counterfeiting to its users by "multi-user-BQC".

## 2.5    The Quantum Computer

> Build quantum computers to simulate nature, because goddammit the world is quantum!
>
> Richard Feynman, Nobel Prize in Physics 1965

The performance and thus the relevance of the future quantum internet will depend not only on its extraordinary security and coordination ability but also on the capabilities quantum computers will develop one day. We will now take a closer look at this extremely innovative concept and highlight the current situation in research.

Already as early as the 1960s, US American physicist Richard Feynman speculated whether quantum physics could be used for computing. It makes a difference whether a computer uses components whose function follows the rules of quantum mechanics (as for example flash memories, TFET transistors, etc.) or whether the calculation process is based on quantum informatics. Only the latter can be seen as true quantum computing. For illustration, let's first take a look at a conventional, classical computer. In simple terms, one might imagine a traditional computer as a kind of black box. This black box is supplied with 0/1 bits (input), and it ultimately again delivers 0/1 bits (output). The actual computing process corresponds to a defined change in the bit sequence, which can be compared to a series of switches that are operated in a predetermined manner. For high processing speeds, many of these switches have to be activated in the shortest possible time. In practice, this is achieved with the help of tiny semiconductor transistors, billions of which are housed on tiny microchips. The sequence in which these switches are triggered corresponds to the programming steps, i.e. the algorithm. A special feature of classical computers is that the sequence of steps is strictly serial (i.e. individual switches are always moved in succession, i.e. in a certain order). This, however, automatically limits the processing speed.

**Moore's Law**

In order to increase the performance of classical computers, the computer industry has for decades relied on a very simple principle. The number of switches is increased continually while the design is made ever more compact. In this way, processor clock rates are improved, making it possible to process an increasing number of bits. At the same time, the power consumption per switch is reduced. In this way, processor performance values per chip have been doubled about every 18 months for many years. This exponential relationship is commonly known as Moore's Law (named after Intel co-founder Gordon E. Moore). Note that this is not a natural law, but rather an ambitious demand set by and for engineers in the computer industry. On the other hand, the fact that such arbitrary scaling is possible at all is rooted in nature's laws. The "silicon revolution" is based on the ability of UV light to etch billions of small transistors out of a thumb-sized silicon wafer. Since UV light has a minimum wavelength of about 10 nm, transistors with an atomic diameter of up to about 30 can be produced in this way. Nevertheless, this game of perpetual miniaturization cannot be continued infinitely, for several reasons. For example, the heat build-up generated by high-performance chips increases much faster than cooling mechanisms can be developed. Another reason is that the wavelengths required for even smaller structures are in the range of X-rays, which cannot be focused sufficiently. Despite these difficulties, structures of no more than 7 nm in size (about 20 atomic diameters) are to be produced in the early 2020s. The ultimate limit, however, establishes an impassable threshold set by the laws of quantum mechanics.

## Quantum Parallelism

The reason for the above-mentioned limit lies in the Heisenberg uncertainty principle, which states that location and speed of a quantum object can never be known precisely at the same time. As a consequence, the exact position of atoms or electrons is uncertain—they are smeared, in a way, or "blurred". As a further consequence, "leakage currents" occur, where the charged particles "tunnel" through the wafer-thin layers of the chips, resulting in short-circuits. From that point on, classical information processing on the basis of electronically coded bits is no longer possible. To solve this fundamental problem, the computer industry has developed, for example, the parallel computer. There, the computing effort is distributed over several processes. This means that tasks are performed separately and simultaneously. At the end, the results are merged again. However, the delegation of subtasks to several chips can be extremely difficult in some cases, as the additional effort to organize and coordinate that distribution can be considerable. Depending on the problem at hand, the temporal coordination of the partial solutions can prove to be a serious difficulty. To date, no standard procedure exists for this. And that is another reason why the computer industry has been searching for concepts that would enable a groundbreaking technology shift, which in turn would bring about a new era, the so-called post-silicon age. Future developments may lie in the use of biological systems, integration of biological and technical information processing, optical signal processing and new physics-based models. The concept of the quantum computer is one of these new approaches. This revolutionary idea addresses very precisely the problem areas of the classical computer. As mentioned above, one basic problem is the serial sequence of the distinct programming steps. This automatically limits the processing speed. What if it were possible to simply override this "causality"? It is precisely this idea that the quantum computer takes advantage of with its ability to calculate "acausally" using a distinct parallel method. This time reversal invariance can not only affect the order of logic gates (i.e. inputs and outputs can be swapped), it also denotes the so-called quantum parallelism. Thus, it becomes possible that several solutions of a problem are already contained in the entire quantum state at the same time and the desired result is read out by a special series of measurements. The main foundation of this strange parallel world lies in both the principle of superposition and its close relative, entanglement. In a manner of speaking, 0 and 1 exist simultaneously, in a multidimensional and strongly correlated form. According to these completely new properties, the bit, the smallest unit of information processing, becomes the qubit of a quantum mechanical linear combination of 0 and 1.

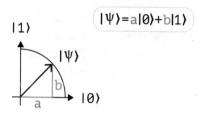

Fig. 2.4 Representation of the linear combination of basis vectors

## 2.5.1 The Qubit—A Multitasking Genius

In order to better understand the properties of qubits, we use a simplified representation. Imagine a circular quadrant bounded by two coordinate axes (Fig. 2.4[2]). Every point on this quarter circle stands for a possible state that the qubit may assume. Each state is represented by an arrow (vector) whose shaft is situated in the origin of the coordinate system and whose apex is directed at the point. If we now imagine the vector moving along the quadrant arc, we get the idea that a qubit unites all these states simultaneously. The number of states is infinite, as already the quarter circle consists of an infinite set of points (and the full circle all the more). Professionally speaking, each individual state is a two-dimensional vector that can be represented as a linear combination of the basis vectors $|0\rangle$ and $|1\rangle$: $|\psi\rangle = a|0\rangle + b|1\rangle$, where the factors $a$ and $b$ stand for arbitrary numbers between 0 and 1. They regulate the length of the basis vectors, so to speak. With the state vector, they always form a right-angled triangle

---

[2]*Note to* Fig. 2.4: This is a simplified representation for didactic purposes. The state of a single qubit is described in QM as a standardized vector in a complex Hilbert space. The states can be depicted as points on the surface of a sphere (Bloch sphere).

whose hypotenuse length is 1. A special case arises for $a$ or $b$ equals 0. Here the triangle has degenerated. The symbol $|\psi\rangle$ (pronounced as "psi vector") is the bra-ket notation of quantum states that is typical for quantum mechanics.

## Nobody is Perfect

In view of the fact that the qubit contains an infinite number of states at the same time, one might reach the misleading conclusion that a single qubit is sufficient for calculations in quantum computers. It is, however, not that simple. The state superposition in a qubit exists only as long as it is not "observed", i.e. in physics terms, as long as the quantum state is not subjected to a measurement. As soon as a measurement occurs, the quantum state decays into one of its two eigenvalues 0 or 1. Another difficulty arises from the fact that it is not possible for quantum states to predict exactly whether 0 or 1 is measured. All that can be provided is a probability, which incidentally can be read from our qubit quadrant. It results from the squares of the factors $a$ respectively $b$. Of course, according to the Pythagorean theorem, the following has to apply: $a^2 + b^2 = 1$, whereby 1 stands for a probability of 100%. If, for example, we assume that the state vector includes an angle of 45° with the coordinate axes, then factors $a$ and $b$ have the same value. Accordingly, 0 or 1 are measured with 50% probability each, since the sum of both probabilities must be 100%. If the angle is 0° or 90° (degenerate triangle), then the probability to measure either 0 or 1 is 100% (because then either $a = 1$ or $b = 1$). In all other cases, different probabilities arise for the measurement of 0 or 1 respectively, depending on the angle. Involuntarily, the question pops up what the advantages of the qubit over the conventional bit would be. After all, its considerable flaw is that while theoretically, it has the capacity to carry an infinite amount of information, it always collapses into its eigenvalues 0 or 1 respectively upon measurement. What's more, this happens with variable probabilities. So at first glance, is seems like the quantum bit might turn out be a boomerang.

## Many Qubits in the Register

The situation suddenly changes when several qubits are brought into superposition with each other—in the best case, an extremely high number. First, imagine a register consisting of two qubits. Already in this case, we have four basis vectors: $|00\rangle$, $|01\rangle$, $|10\rangle$ and $|11\rangle$. Due to the quantum mechanical principle of superposition, any imaginable superposition state can be established from this. We humans aren't able to visualize this, as we are only able to imagine in three dimensions and with this; we are in a four-dimensional coordinate system. We are at least able to express the quantum state of the 2-qubit system in symbolic terms: $|\psi\rangle = a|00\rangle + b|01\rangle + c|10\rangle + d|11\rangle$. By analogy, the corresponding probabilities result from the factors $a$, $b$, $c$ and $d$ whose square sum must be 1. We therefore see that the number of possible linear combinations and thus the number of superposition states has clearly increased. We know that simply because we now (mentally) span a four-dimensional space with our four basic states.

Now, let's consider 3, 4, 5 or a general number of $N$ qubits in the register. We realize that the number of basis vectors increases exponentially, i.e. our state space becomes 8-dimensional, 16-dimensional, 32-dimensional or in general terms, $2^N$-dimensional. If we now compare this fact with a "normal" computer, it becomes evident that with any increase in the number of qubits contained in the register, the possibilities of a quantum computer exceed those of a classical computer by far. The latter requires exactly $N$ bits of information for the representation of a $N$-register, each of which can only stand for 0 or 1. A quantum computer, however, has an incomparably higher number of superposition states, namely $2^N$. In a manner of speaking, that's 0 and 1 simultaneously, and in multidimensional form. It is important to note that not every superposition state can be used for an algorithm. If this were the case, an extremely large number of steps which would contribute to the solution of a problem would be contained in a single state. This leads to considerable reductions in computing time, especially in case of problems with exponentially increasing numbers of calculation steps.

## Operating Principle of a Quantum Computer

It is in fact possible to solve uncomplicated tasks with a simple 2-qubit register. Let's look at an example that illustrates the principle. Imagine a computer as a kind of black box, which accepts an input (0 or 1) and outputs a binary number (0 or 1) for each input. The task for the computer is to determine if both output numbers are identical. In a first step, a suitable quantum system is prepared, which maps the state as a superposition of 0 and 1. In a next step, this superposition state is used as input for the black box. In the black box, the input state (which contains both numbers 0 and 1) is transformed in such a way that both output values can be extracted from it. This is done, for example, with the help of a quantum circuit. Since the laws of quantum mechanics stipulate that no more than one single measurement may be made in order to read information from the state, the system has to be transformed skillfully. This is done in such a way that the measurement is certain to result in a value of 0 if the numbers are equal. Otherwise, the measurement is certain to result in a value of 1. When a concluding *single* measurement is performed, the quantum computer has already completed its task and the solution to the task at hand is available. For a classical computer, it would have been necessary to make *two* requests in order to solve the same problem. Consequently, it performs less efficiently than the quantum computer.

A quantum computer therefore completes the same task more rapidly if, as demonstrated in the above example, it is asked the "right" question. Not the exact value of the calculation is of interest, but rather the fact whether or not the output numbers are identical (the numbers themselves remaining unknown). Later, we will examine a concrete simulation that shows in what way a quantum computer actually works more efficiently. In other words, what question is posed is essential for the quantum computer. It is by no means suitable for solving any arbitrary problem. It is capable of performing very specific tasks only. Those tasks, however, are achieved in drastically shorter periods of time than with a classical computer. The strengths of the quantum computer can be found, for example, in the areas of cryptanalysis, logistics and searches in large databases. The quantum computer provides hope first and foremost for problems that require exponentially increasing computing powers, where even supercomputers fail. It becomes apparent that quantum computers are a very different type of computer that will definitely never replace your notebook or tablet.

## 2.5.2 Quantum Software

At the core of traditional computers are their processor chips, which consist of computing registers and logic gates. The job of the former is the calculation with numbers while the latter process logical decisions in programs. An algorithm is generally understood as a sequence of actions to solve a problem. Several distinct steps can be combined into series of commands in higher programming languages ("software"). Basically, every software instruction has to be translated into machine commands, which the computer then processes. In the case of a classical computer, the code of the machine commands is written in bit language, i.e. in a sequence of zeros and ones. Basically, then, computing is nothing more than converting input bits into output bits in the manner determined by the algorithm encoded in bits. Every step can be represented by logical basic circuits (gates), which are based on mathematical rules.

Similarly, it is also possible to specify instructions for quantum computers. Quantum software in the traditional sense does not exist yet. There are, however, already several quantum algorithms in existence today. In this area, a new branch of computer science is being developed: quantum informatics, which describes the effect of algorithms on qubit registers. Since qubits basically represent superposition states and not 0/1 bits, the mathematics describing them differs fundamentally from that of traditional computers. For example, a quantum circuit consists of several quantum gates that are applied in a fixed time sequence to the qubit register, as for example the quantum Fourier transform which is part of the well-known Shor algorithm. This algorithm is performed on a quantum register with $N$ qubits. Each of the $2^N$ basic states is mapped onto a superposition of all basic states. The result behaves like music, so to speak, where the individual timbre of an instrument is composed of fundamental tones and overtones. A further difference results from the principle of objective randomness. Because of this, many quantum algorithms can only be formulated probabilistically, i.e. they only produce results with a certain probability. However, thanks to the law of large numbers, the error can be kept arbitrarily small by repeating the same measurements a number of times.

## Shor Algorithm

From the "quantum music" of the Fourier series follows an algorithm that would once have taught many cryptologists the meaning of fear. What it calculates is the non-trivial divisor of a composite number. It does this in a much shorter period of time than a classical algorithm. As mentioned already, many encryption methods on the internet are based on the factorization of very large prime numbers. The security of the prevalent "RSA coding" is therefore based, for example, on the fact that no efficient algorithm exists that is able to perform this task in polynomial length. For classical computers, it is not possible to compute problems that cannot be solved in polynomial time in a manageable period of time. In simpler terms, it is much more difficult to break down the number of 323 into a product of two prime numbers than it is to calculate 17 times 19. Now, if the numbers that are to be split have 600 digits or more, this turns into a problem even for supercomputers. The reason is that to date, no mathematical method is known for the efficient calculation of very large prime factors. A technically functional quantum computer on which the Shor algorithm could be performed would, however, be able to overcome this golden rule and thereby make RSA coding redundant. The idea was developed by Peter Shor at Bell Laboratories in 1994. The algorithm consists of a classical element and a quantum component. It also works probabilistically, i.e. in some arbitrarily few cases, it delivers no result at all. Already in 2001, IBM proved that such a "baby quantum computer" is able to break down the number 15 into the factors 3 and 5. That may seem modest and put a condescending smile on some cryptographers' faces. What, however, will happen when the baby grows up? Actually, in the public eye, quantum computers are associated mainly with their alleged code-cracking skills. Occasionally, we see them imagined as potential "monsters". However, for the time being, we can give the all-clear on this. Shor's algorithm is not yet applicable for any practically relevant tasks today. Preventive work is already being done on asymmetric cryptosystems, which should also be resistant to quantum computers. This, then, would be the so-called post-quantum cryptography. After all, no one knows exactly what the state of the art might be in years. New technological developments and breakthroughs in science can lead to utterly surprising concepts and technologies. The danger that existing encryption codes may one day be broken by quantum computers in a matter of seconds can never be ruled out completely. According to whistle-blower Edward Snowden, NSA has been working on a quantum computer like this for many years.

**Grover Algorithm**

An important task of information technology is the search in unsorted databases, be it for search engines or be it for optimization and logistics problems. The fastest possible search algorithm usually is the linear search, which for $N$ entries also requires the same order of magnitude in calculation steps. Suppose you are looking for information about a specific person in the phone book. As long as the names are listed in alphabetical order, you will usually get your result quickly. Not so if the names are listed completely arbitrarily. In the worst case, you would have to look at every single name until you got to the person you wanted to find. For $N$ persons, the maximum number of steps is therefore $N$. Needless to say that the greater the number $N$ is, the longer the time span needed for your search. Classical search algorithms require an average effort of $N/2$ for unsorted data. Now, the Grover algorithm significantly improves on this value. It requires only about $\sqrt{N}$ steps for $N$ entries. For $N = 1$ trillion, that would amount to no more than 1 million steps, for $N = 1$ trillion no more than 1 billion etc. What is more, the memory requirement even scales purely logarithmically, which is very difficult to achieve because the logarithm function increases very slowly. This benefit becomes significant for very large $N$ numbers. A disadvantage may be that the Grover algorithm works probabilistically, i.e. it delivers a correct solution with high probability, but not with certainty. However, the error probability can also be reduced by several repetitions.

The effects of the Grover algorithm can be illustrated by imagining a Mikado game. From a large number of Mikado sticks with the same length and diameter, 4 pieces are extracted and placed in the shape of a square on a piece of paper. Now, a number of $N$ dots is drawn into the enclosed area, which represent the number of search entries. If $N$ is very large, the number of dots will fill the square almost completely. Somewhere in this amalgamation of dots is the one that corresponds to the search entry. In the worst case, a classical computer would have to sift through all of the dots before arriving at the correct one. And that could be a while! A quantum computer, however, works in a fundamentally different way. Quantum parallelism enables it to throw the remaining Mikado sticks in a random jumble onto the square. With very high probability, one of the sticks will exactly hit the correct dot. This is because the length of the Mikado stick corresponds exactly with the lateral edge of the square with the area $N$, and so it only needs to sift through a $\sqrt{N}$ number of dots until it has found the one it was looking for. If none of the sticks hits the correct dot, the process is repeated. With every repetition, the probability to find the required solution increases.

## Quantum Simulators
The domain where the advantage of quantum simulators is especially significant is the simulation of other quantum

systems that cannot be calculated in the classical way, for example in the area of materials science. Every solid has a crystalline structure consisting of billions of atoms or molecules. It is an unimaginably complex quantum system. Its electron states are actually described by the Schrödinger equation, which can theoretically be used to predict every atom, every molecule and thus also the behavior of the entire solid. This is true only in theory, though. Every student of physics knows that the solution of the Schrödinger equation for a hydrogen atom in the ground state is already quite complex, even as this is actually the most straightforward problem. More complex states can only be treated approximately, and even that becomes impossible for powerful computers or supercomputers with 50–60 atoms or more. Therefore, new calculation models that are as close to reality as possible are desirable. Richard Feynman developed the basic idea for quantum simulations which will be presented in the following. In 1982, Feynman proposed an "analog quantum computer" that was not based on digital coding, but imitated nature. A completely different system is used for this, its properties however can be transferred analogously to a partial aspect of the system that is to be simulated. These properties can then be examined without having to simulate the system itself (with the associated savings in storage space and processing power). In this way, the quantum simulator stores and processes the information independently, so to speak, which makes it a quantum computer. Quantum simulators could, for example, predict the behavior of superconductors much more efficiently. One day, they might even be able to reveal the still unexplained mechanism of high-temperature superconductivity. Similarly, the study of quantum magnetism could lead to an increase in the reading and writing speed on the hard disk of a classical computer. Feynman's approach always makes sense when quantum mechanical effects play a major role in a system. As Feynman and others were able to demonstrate, the simulation of such a system on a classical Turing machine would require an exponentially high number of calculation steps, drowning it, so to speak, in its own complexity. Accordingly, classical computing does not represent a solution for such problems. Another major advantage of quantum simulation is that no control is required for each individual component. It therefore offers a significant advantage over other quantum information concepts. A distinction must be made between static models, where the static properties of interacting quantum multi-particle systems are investigated, in contrast to dynamic quantum simulators, which move away out of an equilibrium state and exhibit complex time evolution. Simulations can be carried out in different ways: Digital quantum simulators are based on quantum circuits (see below), which can also be implemented on a quantum computer. Analog simulators are particularly promising because they can, for example, predict the time evolution of an interacting multi-particle system. Already with current technologies, systems have been realized that eclipse even supercomputers. Different physical resources such as ultracold atomic gases or trapped ions are used, but also cavity QED systems or photon condensates. A key objective in this emerging new area is the development of different platforms with high controllability and complexity.

## 2.5.3  Quantum Logic Gates

Classical computers use logical circuits from one or more transistors for the mechanical implementation of their algorithms. Today's microprocessors consist of billions of logic gates. Typically, the XOR gate is used (which relies on the

| bit 1 | bit 2 | XOR | output |
|-------|-------|-----|--------|
| 0 | 0 | → | 0 |
| 1 | 0 | → | 1 |
| 0 | 1 | → | 1 |
| 1 | 1 | → | 0 |

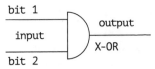

**Fig. 2.5** Classic XOR gate

| target-qubit | control-qubit | CNOT | target-qubit | control-qubit |
|------|------|------|------|------|
| 0 | 0 | ↔ | 0 | 0 |
| 1 | 0 | ↔ | 1 | 0 |
| 0 | 1 | ↔ | 1 | 1 |
| 1 | 1 | ↔ | 0 | 1 |

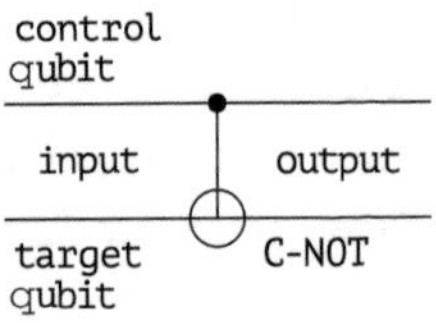

**Fig. 2.6** Qubit CNOT gate

or-function exclusively). It corresponds to a bitwise addition of both inputs modulo 2. This means that whenever the inputs at the gate are assigned unequal bits, the gate generates a logical 1 ("true"), if the bits are equal, it generates a logical 0 ("false"). This can be symbolically represented by a circuit diagram or a truth table (Fig. 2.5).

The CNOT quantum logic gate (Fig. 2.6) extends the characteristics of the XOR further. It has two inputs and two outputs. The truth table shows that the state of the first qubit (control bit) remains unchanged after the gate operation, whereas the value of the second qubit (target bit) follows the logic of the classic XOR. It is inverted exactly when the control bit is set to 1. Likewise, the "acausal" character emerges clearly. The operation behaves in a completely identical manner when run in a forwards or a backwards direction. This time reversal invariance is typical for computational processes in a quantum computer. Another important difference is that superposition states can be used as control and target bits, and special correlated states can be established on their basis.

## Generation of Entangled States

As we have seen, quantum computers profit from quantum parallelism, which is first and foremost expressed in the almost inexhaustible linear combinations of the superposition principle. However, a further phenomenon exists that covers this concept: Einstein's "quantum spook", i.e. entanglement, which is essential for quantum computing. Let's have another look at Fig. 2.6. If, for example, the control bit at the input is set to the superposition state $|0\rangle + |1\rangle$ and the target bit to $|0\rangle$, the CNOT gate generates an entangled Bell state (without normalization factor):

$$\phi = |0\rangle_{control} |0\rangle_{target} + |1\rangle_{control} |1\rangle_{target}.$$

Accordingly, the qubits are entangled with each other via the gate function. An entangled state is characterized by the fact that it cannot be composed of individual partial states of the system components. Rather, its state is created completely anew, and cannot therefore be factorized. In technical jargon it is "not of product form" because it cannot be written as a tensor product of individual qubit states. Incidentally, the superposition state (input control bit) is generated using a single Hadamard gate, which is also able to generate superpositions of all qubits in a register. If we consider two or more qubits to be a single quantum state, that state corresponds to the tensor product of the individual qubits in the register. This is the formal difference between qubits in superposition (which can be written as a tensor product) and entangled qubits (which cannot be written as a tensor product).

**Universal Quantum Gates**

In general it should be noted that quantum computers, unlike their classical counterparts, are not universally programmable. With conventional computers, arbitrary circuits can be constructed from a few basic gates (for example, from a suitable circuit of NAND gates). For quantum computers, a similar kind of disassembly into universal gates is also possible. A set (or "family") of quantum gates is referred to as universal if any (unitary) transformation can be represented as a circuit with quantum gates. It can be shown that a CNOT gate in connection with single-qubit gates is universal. An important example for a single gate is the above-mentioned Hadamard gate (also known as Hadamard transformation H). It generates a superposition state from the base vectors $|0\rangle$ and $|1\rangle$, i.e. either $|0\rangle \rightarrow H \rightarrow |0\rangle + |1\rangle$ or $|1\rangle \rightarrow H \rightarrow |0\rangle + |1\rangle$, each with a probability of 50%. In this representation, normalization factors are omitted that occur due to the condition that the probability in the square sum is 1 (100%). Other examples of single gates are Pauli gates, square roots of NOT gates or the "family" of phase shift gates.

Further entangled states can be created by combining CNOT with single-qubit gates, making this interpretation universal. A "universal" quantum computer is therefore feasible in theory. This is very interesting because it implies that any transformation permitted by quantum mechanics can theoretically be implemented on a quantum computer. With appropriate scaling, it is not only possible to run any quantum program in this way, the simulation of many physical systems also becomes viable. Some experts even believe that a quantum computer would be able to map the entire universe. Already a 300-qubit computer would be as powerful as a computer that uses every atom in the visible universe as its memory cell.

## How Does a Quantum Computer Calculate?

We now take a look at the simulation of a quantum circuit based on the Deutsch algorithm (named after the Israeli-British physicist David Deutsch). The example chosen may seem trivial at first glance, but let's not forget that even conventional computers are only able to calculate at such exceptionally high speeds because so very many operations are executed in extremely short periods of time. As mentioned above, a classical computer can also turn into a lame duck whenever the number of operations increases exponentially. This is why new concepts are needed. For our task, function f is given as $(0, 1) \to (0, 1)$ with the following instruction to the computer: Form the sum f(0) + f(1)modulo2 (i.e. $0 + 0 = 0$, $0 + 1 = 1$, $1 + 0 = 1$ and $1 + 1 = 0$). The input numbers, however, are unknown at the beginning. The computer has to retrieve them first. Note that a classical computer has to call the functions *twice* in order to calculate the results. This is because knowledge of one function value does not automatically lead to knowledge of the other.

Now, we set the same task for a quantum computer. To calculate the result, the quantum computer needs to call the function only once, as we will see in the following. It is therefore in principle able to solve the problem in half the time. Let's take a 2-qubit processor as the simplest option. The basis states at the two inputs are set to the typical input state for quantum computers, $|0\rangle$. What now occurs in the black box can be represented with the help of a quantum circuit (Fig. 2.7), similar to classical computers. Here, N stands for the NOT gate, H for the Hadamard transformation and $U_f$ for the function call.

As can be seen from the table (Fig. 2.4d), the measured values correspond exactly to the solutions of the set task, namely the addition modulo 2. What is revolutionary about this is that all function values are contained "simultaneously" in the superposition state by a single function call. Although

| input | function | | output | measure ment | addition |
|---|---|---|---|---|---|
| $\lvert 0\rangle\lvert 0\rangle$ | $f(0)=0$ | $f(1)=0$ | $\lvert 0\rangle\lvert 1\rangle$ | 0 | 0 + 0 = 0 |
| $\lvert 0\rangle\lvert 0\rangle$ | $f(0)=1$ | $f(1)=1$ | $-\lvert 0\rangle\lvert 1\rangle$ | 0 | 1 + 1 = 0 |
| $\lvert 0\rangle\lvert 0\rangle$ | $f(0)=0$ | $f(1)=1$ | $\lvert 1\rangle\lvert 1\rangle$ | 1 | 0 + 1 = 1 |
| $\lvert 0\rangle\lvert 0\rangle$ | $f(0)=1$ | $f(1)=0$ | $-\lvert 1\rangle\lvert 1\rangle$ | 1 | 1 + 0 = 1 |

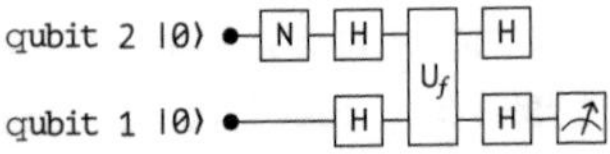

**Fig. 2.7** Quantum circuit and simulation

the laws of quantum mechanics do not permit exact knowledge of the function values, they do allow the calculation of the function regulation. This example illustrates the basic principle that a quantum computer requires the "appropriate" task in order to fully develop its capabilities. On the whole, this example may not seem all that momentous. It can, however, result in considerable acceleration for certain very complex applications where there is an exponential increase in computing effort, because the necessary calculation time is reduced by several billion times.

## 2.5.4 Concepts

Quantum computers replace traditional transistors by qubits, which are implemented with the help of physical resources (such as electron spin or supercurrents). The exponential increase in the computing space can be described by a central wave function, which is represented by a superposition of all possible classical states. It is only through the measurement process that a concrete state is generated, the probability of which is given by the amplitude square of the wave function. In contrast to classical computing, which is divided among various processor cores, probability amplitudes have to be aligned constructively in order to provide a solution to the problem at hand. Successful quantum computing is therefore occasionally compared to a finely tuned orchestra where a (classical) conductor controls tempo and phrasing.

According to the specifications, different concepts for the realization exist. We will briefly address three of them here.

## Circuit Model

This algorithmic model strongly resembles a classical computer (see simulation above) where a quantum program consisting of a quantum circuit and one or more final measurements must be completed in order to obtain calculation results. The results of the measurements can be interpreted as probabilities; the "program" may have to be run several times in order to verify the results. A quantum circuit consists of several quantum logic gates which are applied to the quantum register in a fixed time sequence. Although individual operations can be performed reversibly (as is the case with CNOT), this does not apply to the entire algorithm, which follows a classical sequence of instructions. As mentioned above, all operations are usually limited to single and 2-qubit gates. These are applied to a set of qubits, whereupon the results are read out and displayed at the end as output of a single-qubit measurement. In concrete terms, each basis state is first prepared for computation. In the black box, a universal family of quantum gates acts on the qubits in the desired way. The quantum algorithms are performed by applying single and 2-qubit gates on the computational basis (black box). As explained above, gates of the types H, CNOT, Pauli and phase gates, etc. are usually used (in some cases, also multi-qubit gates). This builds up the necessary degree of entanglement, which causes the speed-up of the quantum computer. If any auxiliary states are used, their states must be cleared so that they do not interfere with the remaining qubits while the computation is performed. The last measurement at the computational basis provides the result. Again, it is important to note that quantum gates represent neither technical nor electronic components, but rather consist of the physical manipulation of one or more qubits. The type of manipulation depends on the implementation method. Usually, excitation states of atoms are influenced by laser pulses, while nuclear or electron spins are influenced by magnetic fields.

## One-Way Quantum Computer

An equally powerful concept that, unlike the circuit model, has no counterpart in classical informatics, is based on measurement-based computation. Typically, the first step is to establish a universal quantum state (usually a strongly entangled "multi-particle state"). Then, the calculation is performed by a series of targeted measurements of this state. The results of previous measurements determine what further measurements are necessary. Thus, the entire resource is provided at the beginning, and the information is then obtained through a series of adaptive single-qubit measurements. The principle can be compared to the legendary tower that contains all the books ever written, so to speak. One then selects from this myriad of books only the infinitesimally small subset that is of personal interest. The name is based on the fact that right at the beginning, as large a number of entangled qubits as possible is provided. The computation is then performed by measuring individual qubits, whereby the entanglement of the output state can no longer be reversed (hence also "one-way"). This is in contrast to the circuit model, where the entanglement is increased successively by applying corresponding gate operations. To obtain the result, it is necessary to employ classical computers in parallel. This is also because the measurement basis sequentially depends on previous results. This measurement-based system also forms the basis for blind quantum computing (BQC) and thus for an inherently secure quantum cloud. Based on the one-way quantum computer, Stephanie Barz was able to implement blind quantum computing in a proof-of-concept as early as 2012.

For a deeper understanding, let's take a look at another example. According to the principle of parametric down conversion, a special laser generates two entangled EPR pairs (i.e. 4 photons, two of which are entangled in polarization). The EPR pairs are then brought into pairwise entanglement using polarizing beam splitters. This system can be regarded as a one-way quantum computer which is given the following task. There are 4 bits in a register, of which three have the value 0 and one has the value 1. Where in the register is 1? While a classical computer has to check an average of 2.25 bits for this task, a quantum computer needs to perform a single step only. The probability for the correct solution is 90%, i.e. the quantum computer delivers the correct result in 9 out of 10 attempts. Of course, this is no more than a proof-of-principle. Any technical implementation of this kind would have to provide a cluster of at least one hundred qubits, which would be almost impossible in experiment.

## Adiabatic Processor ("Quantum Annealing")

Precisely because a huge qubit cluster is very difficult to realize technically, scientists have been looking for a "natural" resource. This is where the idea of a quantum simulator comes into play.

One approach in that direction lies at the heart of the adiabatic quantum computer. There, the ground state of a quantum mechanical system is converted sufficiently slowly into another, more easily readable state. The word adiabatic originates in the field of thermodynamics, where it refers to a system that does not exchange heat with its surroundings. Applied to quantum computers, it describes a physical system where a quantum state of interest is not lost. According to the laws of quantum mechanics, a quantum mechanical system (in its ground state) remains in its ground state even if it changes, as long as this change is sufficiently slow (adiabatic). The idea now is to map the solution of a problem to an initially unknown quantum mechanical ground state of a system. A second system, which is much easier to prepare, is then produced and transferred adiabatically to the first system. If this happens sufficiently slowly, the ground state of interest remains intact and can subsequently be measured. The initial state changes adiabatically to the target state in which the solution is coded. In this interpretation, classical computers are required to "decode" the state that contains the problem. This design was made popular by a Canadian start-up that has been offering commercial applications based on this principle for years. The term "quantum annealing" is, so to speak, the quantum version of the more familiar term simulated annealing. It refers to a heuristic approximation method used to determine a rough solution if the problem at hand is too complex. However, the concept is subject to the restriction that the adiabatic computer in its typical operating condition is not a universal quantum computer; it is mainly used to solve optimization problems.

## 2.5.5 Implementations (Examples)

We have arrived at a basic understanding of the fundamental principles of quantum computing. We now come to the much more difficult question of concrete realizations. To achieve this, we have to radically reexamine and revise what we have learned so far. While in a classical computer, the smallest unit of information, the bit, is usually represented by a voltage value that either exceeds (1) or is below (0) a certain level value, the quantum bit has to be represented by a linear combination of the basis vectors $|0\rangle$ and $|1\rangle$. As mentioned above, a single qubit does not mean anything for the quantum computer. The key question therefore is the following. How can very large quantum registers be achieved, which on the one hand are suitable as storage media and on the other hand can be manipulated and subjected to measurements? After all, it has to be possible to create extremely complex superpositions or entangled states. The technical challenges are daunting, as any interaction with the environment amounts to an unwanted measurement that compromises or destroys quantum coherence. This requires an almost perfect shielding and the best possible stability of the extremely fragile qubits. Similar to the analog computer, these are systems with continuous amplitudes (wave function!), which are far more prone to error than their conventional digital counterparts.

## Relaxation and Decoherence

A fundamental difficulty for the realization of quantum computers lies in the phenomena of relaxation, decoherence and fault tolerance. Closely related to this is the search for a possible architecture and in particular for a concept suitable for scaling (increase in qubit numbers). A general problem here is fault-tolerant computing, i.e. unavoidable errors may not falsify the result in an impermissible way, regardless of the number and distance of the qubits. In addition, the so-called decoherence times have to be much longer than the times required for the gate operations, so that correction by error coding becomes possible. The term relaxation time refers to the characteristic time period until a prepared quantum system enters its stationary state. We all know this problem from everyday life, when our refreshingly cold beer gradually assumes the warmer ambient temperature (and then no longer tastes very good), always striving for thermal equilibrium. Likewise, for a qubit, this leads to it leaping out of its state $|1\rangle$ into the state $|0\rangle$ after a certain time has passed. The probability of such an event happening usually increases exponentially over time. A similarly exponential time behavior is usually also exhibited by decoherence, which denotes the loss of superposition of quantum states. A decoherent qubit behaves only like a classical bit and would therefore be useless. It goes without saying that decoherence and relaxation times have to be sufficiently long for quantum computation to be performed reliably at all. In practice, this usually involves tiny fractions of a second in which quantum error corrections have to be carried out in addition, in order to ensure reliability. This presents the next challenge. Classical methods such as redundancy (perfect copying, multiple storage and comparison of data) are completely ruled out according to the no-cloning theorem. However, it is possible to transfer parts of the quantum information of a qubit onto an entangled system of several qubits. To this end, a kind of code can be generated which is then used to cache the partial information of a qubit in the entangled system. A special measurement is then performed on the qubits, which does not interfere with the relevant quantum information and even reveals information about the type of error (syndrome measurement). This can be used to determine if and which qubit was damaged in which way. The measurement also forces the system into a state that facilitates subsequent error correction. The search for scalable, fault-tolerant quantum systems and quantum gates that can be executed in parallel on different qubits remains an important task of current research. The following is a selection of the countless implementations with which this has been attempted to date.

**Ion Computer and Network**

As explained above, qubits can be realized on charged atoms (ions). Such ion qubits can be controlled and manipulated with very high accuracy. Strong results can be achieved with high fidelity in all gates using macroscopic ion traps. This method is limited by a modest scaling factor, i.e. only a few ions can be captured and addressed individually. Another problem is that a chain of trapped ions of about 14 qubits or more in length increasingly behaves like normal bits. In order to increase their performance, a mini quantum internet would be desirable, where small nodes (each consisting of a few qubits) are interconnected. This has already been demonstrated. Using this system, quantum computing is possible as well, and even conceivable for larger architectures. The entanglement of the cells is achieved by mobile photons on the basis of cavity QED, similar to the interface technique discussed earlier. However, the speed of operations is quite slow. Also, a large number of switches is necessary, something that leads to noticeable photon losses. The priority is therefore to create switches with very low "loss". A new approach lies in integrated ion traps, where standard semiconductor processors are used to realize chip traps in the micrometer range. For this purpose, ions with local magnetic fields are trapped on silicon carrier elements in microwave resonators. The results of this are much better than when one manipulates each ion individually, especially as the lasers have to be aligned ultra-precisely. A further advantage is that the cooling capacity can be reduced (to a "mere" 70 K $= -203.15°$ approximately). The process is supported by nitrogen cooling, which takes place in microchannels located above the chip modules. This is also a major advantage over conventional superconductor systems, which require an inordinately greater cooling effort. Another possibility would be to no longer read out the quantum information by laser, but to use specifically adapted computing modules. These are accommodated in tiny modular vacuum cells, whereby specific error correction systems are implemented as well. All in all, this is a promising concept which however requires a large number of qubits to achieve high performance rates. There is still much that needs to be done. Nevertheless, the mood is cautiously optimistic, as proven by numerous start-ups.

## Superconductor Qubits

Generally speaking, a superconductor is a material that undergoes a phase transition when it falls below a specific transition temperature (which typically is very low). After that, it enables currents to flow through it without electrical resistance. Already today, this strange quantum effect has a firm place in many technical applications. These include for example the generation of strong magnetic fields or their highly sensitive measurement. An often-considered future vision concerns the tremendous amounts of electrical energy a superconducting system would save. It is, however, important to consider the amount of energy required to achieve the cooling effect. With this, the advantage is immediately called into question again. Numerous explanatory approaches and models are still being discussed theoretically today. On a practical level, there is a concentrated search for "hot" superconductors (which function at room temperature). Possible applications include the high-voltage lines of future maglev trains, which would then be able to move essentially free of friction. The possibility of superconducting qubits also opens up a significant number of possibilities for quantum computers. One example is the so-called SQUID, a superconducting ring that is interrupted in the center by a very thin insulator. Electron pairs can "tunnel" through this microscopic slit. The qubits in superposition consist of super streams that flow in opposing directions. What would be completely impossible in classical physics, however, has become commonplace in the world of quantum theory. Tests on single-qubit gates already today achieve a fidelity of 99.9%, and even on 2-qubit gates, fidelities of up to 99.4% are reached. In addition, this "fidelity" (measure for the error rate) lies within the tolerance limit for the error code, which enables the realization of certain qubit arrangements. Although this system benefits from the miniature design of its devices, several disadvantages remain. One of them concerns the "crosstalk" between the nanocables. Because of this effect, three-dimensional structures are hardly possible (something that would be of advantage, also in view of fault tolerance). The most significant disadvantage however is that the relevant superconductors only work in ranges approaching absolute zero ($-273.15\,^\circ$C). This poses considerable challenges for scaling, if only because of limited cooling capacities. This application has become known to the public mainly through the "quantum department" of Google.

## Photonic Computing

In earlier approaches, quantum computing concepts based on several photons failed because the methods for producing and manipulating entangled photon pairs were not sufficiently advanced. In the meantime, this has changed. On the basis of so-called coherent photon conversion, it is now possible to bring photons into special interaction with each other—without destroying their quantum information. For this purpose, the light of two single photon lasers of different wavelengths (and thus energy) is introduced simultaneously into a fiber-optic line. The overall impact is that a relevant single photon state can then be converted into a two-photon state. As a rule, photons do not interact at all with their own species, as can be seen from the fact that the light beams of two flashlights can be crossed without obstructing each other. In a sense, photons are isolated, and therefore they are able to carry quantum information very efficiently. This also forms the basis of linear optics, where it is assumed that the optical properties of a material behave independently from the intensity of the incident light. However, in order to use light for quantum computers, interaction between individual photons is necessary. This is made possible by highly nonlinear optical materials whose behavior is influenced by the intensity of the incident laser light. This concept was proposed already years ago by physicists from the University of Vienna, and from Japan and Australia. Not exactly predictable 2-qubit operations and photon losses, however, still pose major challenges for these technologies. Recent theoretical breakthroughs in combination with technical advances have rendered this system a viable competitor in the race for the quantum computer. The architecture uses a similar approach to that of the one-way quantum computer. The ability to miniaturize optical elements on a single chip using nanofabrication technology is also a promising sign for the potential realization of optical quantum computers with millions of elements per chip. The fact that cooling equipment is only required for photon detectors and will in the long run no longer be necessary is a further advantage. Recent research indicates that three-photon states can also be generated in a similar way, which leads to new avenues for the realization of the photon computer.

## 2.5.6 Quantum Supremacy

The concept of "quantum supremacy" is a milestone for the current development of quantum computers. The (somewhat nebulous) term was coined by the American theorist John Preskill. It defines the point where quantum processors will for the first time succeed in surpassing the computing power of even supercomputers (at least regarding specific applications). For years, various research teams, first and foremost high-tech giants such as IBM

and Google, have been working intensively to achieve
this goal. One example of this optimism is John Martinis
from Google Research Labs at the University of California
in St. Barbara. He brought together leading scientists to
demonstrate a prototype with 22 qubits in two rows of 11
qubits each. Next, a quantum chip with 49 qubits went
into operation—in $7 \times 7$ format and based on superconducting qubits. Compared to ion computers, such as those
operated by his colleague Chris Monroe at the University
of Maryland, Martinis' qubits are enormous. The relevant
component consists of a small metal cross about 0.5 mm
in length which is cut from foil. At its ends, there is a
Josephson contact consisting of two superconducting layers with an extremely thin insulator between them, similar
to a SQUID (see above). This "sandwich" is then cooled
down to a temperature near absolute zero. Now its behavior can only be explained by quantum mechanics. As a
result of the quantum mechanical tunnel effect, electrons
can tunnel through the insulation layer, turning the entire
structure into a qubit. This is possible because the current
flows in both directions simultaneously due to the superposition principle and superconductor technology. It is
controlled and read out via microwaves in the GHz range.
This electromagnetic radiation also serves to entangle the
qubits. Admittedly, this is first and foremost a system
test with the aim to bring the quantum computer from
purely fundamental research into a concrete technology.
Only a few of the quantum gates are randomly switched.
This effect by itself does not result in a usable algorithm.
Martinis is optimistic, however. He sees his project as
being on the right track. Of high importance is the quality of the qubits and the associated low error rate. This
facilitates several hundred operations before the qubits fall
victim to decoherence. According to recent media reports,
Google has officially announced having achieved quantum

supremacy. In an article in the science magazine Nature, Google claims to have performed operations with the 54 qubit processor "Syncamore" (of which 53 qbits are actually working) for 200s, operations for which even the supercomputer "Summit" (owned by IBM) would require many thousands of years. It's not surprising that this statement is questioned by Google's competitor IBM (also in possession of a 53 qubit processor). The paper outlining the work demonstrates that the superconducting processor Syncamore was able to perform a sampling random calculation—essentially verifying that a set of numbers is randomly distributed, exponentially faster than any classical computer. IBM has already pushed back on the claim, stating preference for a higher threshold on quantum supremacy, and insisting that with a clever programming a classical machine is able to solve the same problem in just a few days. According to some experts this is, however, hard to jugde and should not detract too much from Google's achievement. Although really powerful quantum computers may remain decades away, this might be an important step on the long road.

**The Race Has Begun**

The run on the quantum computer has long since begun. In addition to mainly US-based computer and IT giants, Europe, but above all China, are important contenders. Companies with big names already today offer access to quantum processors via the internet. This, however, has nothing to do with a "real" quantum cloud in the sense of powerful blind quantum computing. Rather, this is something like a playful getting-to-know of rudimentary aspects and possible applications. The basic idea, however, is directed towards the future. One day, a worldwide community could be involved directly in the development of quantum software. We should not set our expectations too high at the moment, however. Compared to the development of conventional computers, the quantum computer has just reached its infancy. As we know, the performance of the very first computers was light years away from today's standards. On the other hand, today we live in a highly technological time, where development leaps occur within ever shorter periods of time. Worldwide communication via the internet also contributes to this, which means that science is also in a state of constant intercontinental exchange. The central scientific challenges on the way to the quantum computer are as follows. Fidelity should be as close as possible to 100% in order to provide the conditions for efficient error correction procedures. One must bear in mind that an error rate of only 0.1–1% in practical implementations requires roughly 10,000 additional qubits for redundancy. Consequently, systems are necessary that avoid such absurd increases in complexity. Equally important are coherence times and gate speeds. Above all, it is also necessary to improve the initialization and readout qualities and the speeds of the qubits in order to be able to implement meaningful error algorithms. Probably of the highest importance is the search for scalable architectures. This requires the production of corresponding chips, but also the development of suitable control systems for the operations by means of optimized control voltages, also for laser light, radio waves or microwave pulses. After all, the classical hardware and that of the quantum computer have to be optimally matched to each other. Further development goals lie in the compatibility with conventional semiconductor manufacturing processes. At the moment, while experts are hardly willing to risk concrete predictions, they consider a period of 10–20 years until the first real breakthrough to be probable. Once this major step has been achieved, further development will likely proceed very rapidly and trigger developments still beyond the horizon today. Until then, however, it will be necessary to find ways to cope with the enormous technical difficulties that have just been outlined. From a physics point of view, it is particularly interesting to what extent the challenge of decoherence can be met.

The other side of the coin relates to business, i.e. the potential market value of this technology, and consequently global relevance, investment and prestige. The latter certainly plays an important role in countries like China. Even if, however, "full supremacy" is achieved, i.e. a quantum computer outperforms a classical computer in relevant algorithms for the first time, what would be gained? Cost and complexity would limit its significance to a great extent. And, of course, the development of classical computers never stops either. Already today, conventional computers can be used universally, and far more economically. Why would one charter a jumbo jet just to cross the street? Apparently, corporate labs are under a certain amount of pressure. Quantum computers are a commodity of the future, and the stakes are high. Such prospects might deter investors who have already gambled hundreds of millions of dollars on relevant start-ups. Considering the sums that are in the game elsewhere, however, the quantum computer as a strategic investment is certainly a risk worth taking. One day, its value might skyrocket. By the way, the comparison with the classical computer is misleading. The quantum computer will never be a substitute for conventional computing devices. Rather, it's all about special applications that cannot be tackled by classical means yet. It's a fact that starting with about 50 qubits, a classical computer threatens to suffocate in its own complexity, no longer able to sensibly handle certain problems. This is where the quantum computer comes in. Ultimately, it is about the inimitable, about the innovative, about completely new approaches to solutions and aspects that only the quantum computer of the future would be able to deliver. And that's not even mentioning its value for science. John Preskill himself states that quantum supremacy should not be misunderstood or overestimated. According to his assessment, one is at the moment in an initial era ("noisy stage") caused by decoherence. The quantum computer will have a revolutionary effect on humanity, however. In any case, it should be noted that the quantum computer is definitely the "hottest commodity" and the most exciting question of future information technology.

## 2.6 Tap-Proof Data Transmission

It seems that humanity's desire to send secret messages that cannot be read by unauthorized persons has existed for thousands of years. The oldest known evidence can be found in ancient Egyptian cryptography in the third millennium BC. In the Middle Ages, a variety of ciphers was used mainly for diplomatic but also for medical correspondence. Widely known is the encryption machine Enigma, which was used by the Germans in World War II and cracked by allied scientists led by Alan Turing. With Claude Shannon, scientific cryptography emerged as a mathematical discipline after the Second World War. Today's age of comprehensive digitalization and global networking, on the other hand, is associated with the topic of information security, which also includes concepts for resistance to manipulation and unauthorized access. In this, objectives like the protection of data stocks are pursued, but also concepts such as confidentiality, integrity, authenticity or commitment. This includes protective measures against unauthorized access, alteration of data, forgery and intellectual theft.

## 2.6.1  Classical Encryptions

The enormous field of cryptography offers countless possibilities and algorithms. Nevertheless, its processes can be divided into three basic types: symmetrical, asymmetrical and hybrid systems. Symmetrical encryptions are the simplest and use one key per communication partner. This key must not be known to anyone but the two communication partners and may be used only once. The respective document is encrypted with the key and then sent via the internet. The recipient decrypts this ciphertext using the key. The result is plain text. Such systems are easy to use and therefore suitable for large amounts of data that need to be transferred quickly. The security of the system depends, among other things, on the key length. To illustrate the principle, let's take a look at the numbers. If the key is only 1 byte long, i.e. corresponds to a random sequence of 8 binary numbers, then there are $2^8 = 256$ possibilities for the key. This, of course, wouldn't provide very good protection, since a computer using a "brute force attack" would easily be able to check all 256 possibilities and therefore find the right key immediately. If the key is 256 bits long, there are already $2^{256} \approx 10^{77}$ variations. That's almost as many as there are atoms in the entire universe! If, for example, a video conference is encrypted in real time, the bit rate is in the Mbit/s range. The mathematical probability that a hacker attack then leads to the target de facto amounts to zero, and that remains true even if a supercomputer is available. Security therefore depends primarily on the key length (and the randomness of its bit sequence). However, one challenge remains. In practice, the key is always generated algorithmically. Computers are not able to generate real random numbers, they are always "pseudo-random". This makes such systems vulnerable, because it calls into question an important prerequisite, namely that all possible bit sequences have exactly the same probability. The decisive problem of symmetric encryption, however, lies in the key distribution from sender to receiver. In practice, the message is always transmitted via the internet and can therefore be tapped at any point. It goes without saying that the key can be stored at any point without authorization.

Typically, asymmetrical encodings are used to improve security. Examples include the prevalent RSA encoding, which is based on the factorization of very large prime numbers. The procedure is based on a public key and a private key, both of which have to be kept secret. Both parts are mathematically connected in such a way that messages encrypted using the public key can only be decrypted with the corresponding private key. In order to transport a document from A to B, the data is encrypted with the public key and then sent. After that, it can only be opened with the recipient's secret key. For this method to be effective, however, one has to make sure that the public key is correctly assigned to the recipient. It is important that the functions that are used are not reversible (or can only be reverted with extreme difficulty). Otherwise it will be possible to calculate or reconstruct the private key from the public key. In practice, this is very often achieved by so-called one-way functions. However, the public key procedure can also be run in both directions, if one takes into consideration the possibilities of the digital signature. Encryption systems are used to not only safeguard confidentiality and integrity, but also the authenticity and the binding nature of a message. However, the security guaranteed by symmetrical procedures cannot be evaluated equivalently with an asymmetrical mode in terms of information theory, because for a sufficiently powerful attacker it would always be possible to solve the underlying mathematical problem. As mentioned earlier, quantum computers in particular are able to do just that.

Purely asymmetric methods are inherently very slow. That's why hybrid methods are used to combine symmetric and asymmetric encryptions. In such procedures, the document to be transferred is first encrypted symmetrically using a randomly generated electronic key (session key). This system offers a considerable speed advantage, especially if very large data volumes come into play. In the next step, the session key is encrypted asymmetrically using the recipient's public key. The two pieces of information (encrypted document and encrypted session key) are then sent to the recipient in a mutual cryptographic message. To decrypt it, the session key is first decrypted asymmetrically, using the receiver's private key. In this way, the recipient finally receives the session key, enabling them to decrypt the document symmetrically. In fact, practice-oriented variants almost always work with hybrid systems. This is because usually, the amount of user data is rather large, and so the speed advantage of symmetrical encryption is used. On the other hand, the session key usually remains relatively short, so the slow asymmetric encryption does not matter very much, and it minimizes the risks of key distribution.

Hash methods are also often used to generate a unique fingerprint from information of any length. The idea behind this is comparatively simple. The sender generates a hash value (for example, the sum of the digits of the data interpreted as a binary number) from the information to be protected and sends this hash value together with the information to the recipient. The recipient then calculates the hash value of the received data and compares it with the hash value transmitted by the sender. If both values agree, this implies that the information has remained unchanged during the transfer. But this is precisely where the difficulty lies. On the one hand, the hash procedure should be able to manage larger files in a short period of time. On the other hand, the hash value needs to be unambiguous. All this makes it extremely difficult to manipulate the information.

Only a subset of the digital security techniques in existence today are covered by the examples above. In addition, many methods with great development potential are being investigated today. This up-and-coming industry will become more and more important in the future, and the number of IT specialists working will increase with it.

### The One-Time Pad (OTP)

A particularly straightforward method for demonstrating digital cryptography is the symmetrical one-time pad (OTP). A person, party or computer (let's call them Alice) wishes to send encrypted data to another person/party/computer (Bob) over the internet. Alice encrypts her message and sends it publicly over the internet. Bob decodes the received cipher with the secret key known only to him and Alice (private key) and receives the plain text message. Such encryption is performed, for example, by adding binary numbers via the XOR operation, $0 + 0 = 0$, $0 + 1 = 1$, $1 + 0 = 1$, and $1 + 1 = 0$. This means that Alice encodes her message in binary numbers and generates the cipher via binary addition with the private key. Bob adds the key to the ciphertext and receives the plaintext. An example is shown in Fig. 2.8.

Despite its simplicity, this procedure can be regarded as 100% secure, i.e. perfectly secure in terms of information theory, under the following conditions:

1. The key has to be an absolute original and may be used only once. Not even parts of it may be reused.

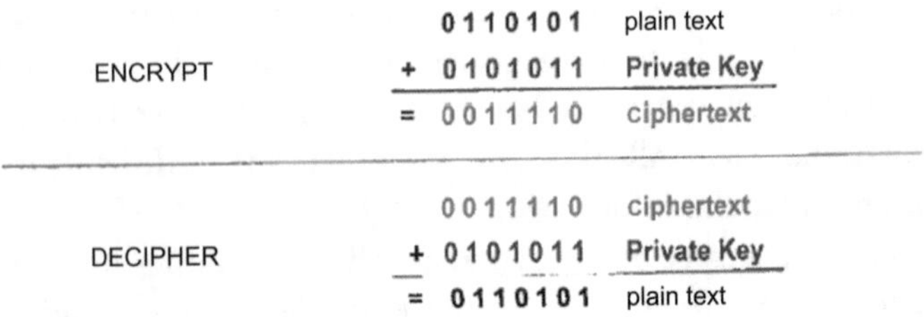

```
                    0 1 1 0 1 0 1    plain text
ENCRYPT         +   0 1 0 1 0 1 1    Private Key
                = 0 0 1 1 1 1 0      ciphertext
```

```
                    0 0 1 1 1 1 0    ciphertext
DECIPHER        +   0 1 0 1 0 1 1    Private Key
                = 0 1 1 0 1 0 1      plain text
```

**Fig. 2.8** One-time pad, example

2. The key length must be at least as long as the data to be sent, so that a brute force attack has no realistic chance to succeed.

3. The key must remain secret, i.e. it must be known only to Alice and Bob.

4. The generated key has an equally random distribution of 100%.

5. Human errors such as mistakes or corruption are ruled out.

OTP therefore fulfills the important principle that security does not depend on keeping the algorithm secret, but rather on keeping the key secret. As mentioned above, in practice, requirement 4, but above all point 3, represent a serious safety concern, which means that a purely symmetrical procedure is no longer appropriate today. In combination with quantum technology, however, this fundamental problem is solved in an ideal way.

## 2.6.2 Quantum Key Distribution (QKD)

A central problem of today's IT is its lack of physical security. This means that classical information can always be reproduced and intercepted at will. It is therefore exposed to unauthorized and criminal use at any time. Even the most ingenious security system then basically relies on the creation of obstacles for the hacker or spy, in order to make it more difficult to get access to the data. In most cases, security is based exclusively on the difficulty of mathematical problems and the confidence that the attacker's computing performance is inadequate. This is precisely where quantum physics comes into play. The term quantum cryptography generally refers to the use of quantum mechanical effects as components of cryptographic processes. Among the several interpretations currently under investigation, the quantum key exchange (QKD) approach should be mentioned above all. It represents a promising solution to the problem of key distribution in symmetric coding. Its objectives are as follows:

1. Alice and Bob agree on a mutual secret private key, without spy Eve (named without ill intent) being able to discover it in any way.
2. An inherently safe quantum channel is created for the transmission of the quantum key. Its security is based on the fact that any eavesdropping attempt by Eve changes the quantum channel in such a way that the hacking attack immediately becomes evident. While Eve may try to copy the quantum state, according to the no-cloning principle, however, it is impossible to copy it perfectly.
3. There has to be a possibility for Alice and Bob to securely authenticate the message, in order to prevent Eve from taking the position of either Alice or Bob. If that is the case, the security of the QKD can also be proven against unlimited attacks, which is not possible with classic key exchange protocols.

**Encryption Protocols**

In general, QKD systems are always two-step processes. In the first stage, a sequence of true random numbers is generated. In the second stage, the generated quantum key is used for classical cryptography and sent via the normal internet. A simple symmetrical procedure such as the one-time pad (OTP) (which is proven to be secure by quantum technology) is sufficient for this. Because the key is generated quantum randomly, i.e. the randomness originates in nature, it is not possible to "recalculate" the key algorithmically. In addition, each bit sequence generated has exactly the same probability and offers (depending on the key length) such an astronomical number of possible combinations that the procedure can be regarded as maximally safe. At least, this is true for any relevant period of time, which even for supercomputers can amount to many millennia and more. There are basically two procedures for the QKD (all variants are based on these).

**BB84 Protocol**

This is the most popular method for quantum cryptography. It was formulated at IBM in 1984 by the two physicists Charles H. Bennett and Gilles Brassard. The basic idea, however, goes back to Stephen J. Wiesner, who suggested it around 1970. BB84 is based on the transmission of single qubits, which are usually implemented as polarizations or phases of photons. The transmission takes place either via fiber-optic cable or in direct line of sight. The majority of the devices that are commercially available and (according to the manufacturer) used by governments and companies, but also strategic investment partners, work with this protocol. In the Swiss parliamentary elections in 2007, results from polling stations in the canton of Geneva were transferred in this way to Bern over a distance of approximately 100 km.

## Ekert Protocol

This technically much more complex, but very promising method was proposed by Artur Ekert in 1991. In stark contrast to BB84, this protocol creates entangled qubits. The first quantum optical realization was achieved in 1999 by Anton Zeilinger and his group over a distance of 360 m. By the same team, the first money transfer via quantum cryptography was demonstrated in 2004, which caused quite a stir in the media. In the presence of Vienna's mayor at the time, Michael Häupl, money was transferred from a bank more than 1.5 km away to the Vienna City Hall. The proverbial quantum leap in this development can be seen in the first intercontinental transmission by quantum satellites, which was performed successfully on 29 September 2017 (Sect. 1.4).

### The Ekert Protocol, Step by Step

*Step 1—Authentication*

First, Alice and Bob have to make sure that Eve won't be able to take the position of either one of them. A number of quantum methods are suitable for this, referred to as "quantum passwords" hereafter. In addition, an authenticated classical channel has to be set up, which can be intercepted.

*Step 2—Quantum Key Generation (QKD)*

From a source that emits entangled particles, specifically correlated single qubits are sent to Alice and Bob, which decay quantum randomly to their eigenvalues of 0 or 1 upon measurement. From a sufficiently long series of measurements, Alice and Bob sift those cases with clear correlations (relevant bits) from those without clear correlations (irrelevant bits). For this they need a classical information channel.

*Step 3—Error Correction*

The measurement errors that are unavoidable in practice are corrected using special procedures (as for example parity test, privacy amplification, etc.).

*Step 4—Spy Test*

Real security can only be guaranteed if the entanglement remains maximally intact. If this is not the case, the system becomes either partially or completely decoherent. This can be recorded statistically, for example by evaluating Bell's inequality. If the inequality is not violated, it can be assumed that there has been an eavesdropping attack, or that there is a technical problem. In this way, any successful espionage attack can be detected immediately while at the same time, the functionality of the system is checked.

*Step 5—OTP Encryption*

Only after the above steps have been completed is the actual data transferred. The quantum key (consisting of the error-corrected relevant bits) is used for a symmetrical procedure such as OTP and sent over the regular internet.

*Step 6—Deciphering*

Bob receives the ciphertext and deciphers it with the private key, which was also created by him. The exceptional thing about QKD is that the key is not transferred from one location to another. Rather, it is created by Alice and Bob via the measuring process "as if by itself". The inherent security of the system is based on the no-cloning theorem, which makes it impossible to intercept qubits.

## 2.6.3 Quantum Cryptography with Entangled Photons

In the following, a concrete test arrangement is presented which demonstrates the method of quantum key distribution (QKD) according to the Ekert protocol in detail (Fig. 2.9). This can be implemented in satellite-based systems (quantum satellite) as well as in fiber-optic networks. The quantum information in this example is coded in its polarization (vibrational plane) of light. In principle, however, it can also be inscribed via the phase or the energy-time uncertainty. Alice and Bob generally stand for two persons/parties/computers

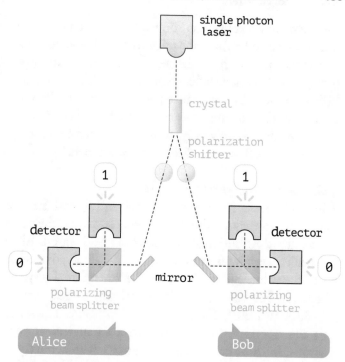

**Fig. 2.9** Experimental setup for the Ekert protocol. The entanglement source is a nonlinear *crystal*. The degree of entanglement can be influenced by additional elements that are not depicted here. The controllable *polarization shifter* changes the vibrational plane of the linearly polarized light according to its setting. The *polarizing beam splitter* splits the laser light into horizontally and vertically polarized light and with the *detector*, single photons are measured

who want to send each other inherently tap-proof digital information. In quantum satellites (such as QUESS), Alice and Bob correspond to observational measuring equipment.

## Experimental Setup

The entanglement source ES generates a large number of maximally entangled photon pairs according to the principle of parametric down conversion (dotted lines). One partner particle is measured by Alice, the other by Bob. Before the measurement is performed by the detectors D, the photons pass through a polarization shifter PS and a polarizing beam splitter PBS. Depending on the relative positions of PS and PBS, the photons in the PBS are either transmitted or reflected. These directions are arbitrarily assigned the binary numbers 0 or 1. In our example, 0 = transmitted and 1 = reflected.

## Collection of Measurement Data

After each generation of a particle pair in ES, both PSs are rotated completely randomly and independently (the same probability distribution should apply in both cases). Alice sets her PS exactly to the values 0° or 30°, Bob adjusts his PS to the values 30° or 60°. There are therefore 4 ways in which the PSs can be positioned relative to each other: (0°, 30°), (0°, 60°), (30°, 60°) and (30°, 30°). However, only the case (30°, 30°) yields perfect entanglement correlations. In those cases, Alice measures 1 with certainty if Bob's measurement also results in 1, and Alice receives the measurement result of 0 whenever Bob gets a 0. In these cases, one speaks of *relevant bits*. However, whether the result is actually 0 or 1 is determined purely by objective quantum randomness (Sect. 2.6.3 "Further details"). For all other pairs of angles, there are no perfect correlations, i.e. it cannot be confirmed with certainty whether Alice and Bob measure the same binary numbers. These remaining cases are the *irrelevant bits*. Each individual measurement result is recorded in numerical order so that Alice and Bob each receive a list of relevant and irrelevant bits.

## Testing for Eavesdropping Attempts

When the number of error-corrected bits on the list has reached sufficient quantities, a statistical evaluation is performed. Based on this, probabilities can be calculated which either violate the Bell inequality or not. If it is violated, this proves that the entanglement was completely intact during the bit generation and that the quantum channel can be considered to be safe. If, on the other hand, the Bell inequality is fulfilled (not violated), the quantum channel has been corrupted, either by an eavesdropping attack or due to a technical defect, and the transmission must be regarded as insecure. The Bell inequality can be formulated in many ways for various practical interpretations, as for example the popular CHSH inequality. In our example, we present an inequality according to Eugene Wigner. In relation to the specified angle or Bell state $\Phi_+$ (= specially adjusted entanglement state), the Wigner inequality is written as follows.

$$P_{++}\left(0°, 30°\right) < P_{++}\left(0°, 60°\right) + P_{+-}\left(0°, 30°\right)$$

where $P$ denotes the probability that certain bit combinations for the specified angle pairs will be measured for Alice and Bob. The symbol "++" indicates cases where a perfect correlation is given. An estimate for $P$ is the relative frequency of events. Due to the law of large numbers, the estimation is arbitrarily reliable after a sufficient number of measured values has been reached. If, for example, the Wigner inequality results in $0.35 < 0.13 + 0.13$, then it is obviously violated, since $0.35 < 0.26$ is false. In this case it can be concluded that the quantum channel was not attacked.

**Quantum Key**

The Wigner inequality in the above example makes successful eavesdropping attacks impossible. And so, in the next step, Alice and Bob separate the relevant bits from the irrelevant ones. The set of error-corrected relevant bits is their result: the finished quantum key. In contrast to classical IT, the key

1. was generated directly by Alice and Bob without being transmitted via the internet,
2. is an absolute original every time, because of quantum randomness and
3. is a truly random sequence, in contrast to pseudo-random numbers generated by computer algorithms.

   Because objective randomness is an irreducible event (as the Bell theorem proves indirectly), quantum random numbers are the best there can be.

It should be noted that the real random sequence is not generated during the (possibly not completely random) rotation of the PS, but at the PBS (Sect. 2.6.3, "More Details").

**Use for OTP**
See Sect. 2.6.1.

**Inherent Safety**
Again, Eve is an unauthorized person/party/computer who wants to eavesdrop on the data to be transmitted. Eve basically has two options. She can attack either the classical channel or the quantum channel.

**Attack on the Classical Channel**
To generate the quantum key, Alice and Bob require a classical information channel, that is, a conventional IT connection. At first glance, this seems to be a worthwhile target for all kinds of espionage attacks. It should however be noted that Alice and Bob only exchange a list of angle specifications rather than concrete bit sequences. This information is sufficient for both, especially since they know that whenever their angles are identical, they get perfect correlations. By changing their detector settings, they can determine whether this is 0 or 1. Eve, on the

other hand, can never know the values of the relevant bits because they are objectively random with Alice and Bob. Even if Eve had the information that perfect correlations occur at (30°, 30°), it is always completely unclear every time whether these are zeros or ones. Finally, Eve has the option of hacking into the OTP encrypted cipher and trying to crack it using a brute force attack. However, since the QKD meets all security criteria for the use of OTP (which is doubtful in the case of a typical IT key allocation), this endeavor will inevitably fail as well.

## Attack on the Quantum Channel

In principle, Eve has three options for attacking the quantum channel.

*Attack 1*: Eve monitors the direct line of sight (or fiber-optic link) between the entanglement source ES and the detectors. To do this, however, she has to measure the polarization state of the photons, which automatically disturbs the entire quantum state and changes the measurement statistics so that the Wigner inequality is no longer violated. Of course, Alice and Bob will notice this immediately and ignore the resulting quantum key.

*Attack 2*: Eve listens directly to the ES and, for example, transmits the resulting quantum key to a secret location. This, however, is also impossible, because the bits are created objectively randomly during the measuring process with Alice and Bob. They are not created in the ES. A measurement attempt by Eve would ultimately result in attack 1 again. Such attacks would only be successful if it were possible that the entire quantum state also exists with Eve, independently of Alice and Bob. This would require at least a (perfect) doubling of the quantum information, which is physically impossible due to the no-cloning theorem (Sect. 2.1). Of course, it is conceivable that Alice and Bob's measuring devices secretly transmit the generated quantum key and store it on a classical computer. Then, of course, Alice and Bob have to make sure that their devices are not manipulated in any way. They have to interpret the inequality correctly.

*Attack 3*: Eve starts a man-in-the-middle attack. To achieve that, she poses as the source and simulates the entangled photon pairs. Alice and Bob might not become aware of this, but even in this case, the Wigner inequality is no longer violated. The reason is that Eve would only be able to do this in a classical physical way, so there is no entanglement per se. However, since the Wigner inequality is a direct criterion for the existence of the entanglement, this deception maneuver automatically becomes conspicuous as well.

## Further Details

### Parametric Down Conversion

In this example, the single photon source generates photons with a wavelength of 405 nm (blue light). The light quanta meet a nonlinear crystal system, whereupon an entangled pair of infrared photons of double wavelength 810 nm is formed from each blue photon. The photon pairs generated in this way then move along a cone, i.e. with a certain probability they can describe all those directions that correspond to a cone shell. To generate the entanglement, it is necessary to place two nonlinear crystals one behind the other in such a way that two such cone shells are formed, which partially overlap each other. Along these overlapping areas, the particles are indistinguishable (i.e. there is a kind of loss of information), which is an important prerequisite for entanglement. The degree of entanglement can be adjusted by means of additional elements such as polarization rotators and phase shifters. In the example given, four different maximum entangled states (so-called Bell states) with different polarizations can be generated.

## Malus's Law

Whenever polarized light passes through an analyzer, its polarization plane is rotated according to the position of the filter. This shows that the intensity (brightness) of the transmitted light decreases more and more as the rotation angle increases, until finally, nothing at all is transmitted at 90°. According to Malus' law (see also Sect. 1.5) this decrease is proportional to the squared cosine ($\cos^2$) of the angle. If, on the other hand, polarized light passes through a polarizing beam splitter, one half of the light is transmitted at an angle of 45° and the other half is reflected. At other angles, other ratios are formed. At an angle of 30°, a ratio of 75–25% occurs. From a quantum perspective, individual light quanta pass through the beam splitter in these cases. Whether the individual photon is transmitted or reflected is objectively random. Although the overall statistics exhibit a percentage distribution as given above, the individual quantum mechanical event is subject to absolute randomness. Making sure that that the angles are not 0° or 90° is important, as otherwise the photons are either transmitted or reflected to 100% and the arrangement then no longer functions as a quantum random number generator.

## The Photon as Qubit

Let's return to the schematic of the linear combination on the quarter circle in Sect. 2.5.1 (Fig. 2.10). We will expand this idea to a full circle. To begin with, this does not change the factors $a$ and $b$ in front of the basis vectors, except that they now may also have negative values. This is why for the probability, the condition has to be set that the absolute square values of the sum must equal 1 (note: factors $a$ and $b$ in general are complex numbers). A qubit

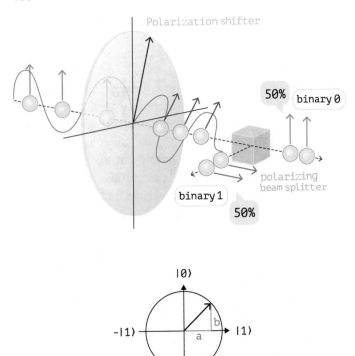

**Fig. 2.10** Representation and measurement of a qubit (implemented as a photon)

visualized in this way can now be associated directly with the polarization of a photon. Just as there exist an infinite number of states of a qubit, there are theoretically just as many infinite possibilities for measuring the polarization state of a photon. The absolute squares of the factors $a$ and $b$ then indicate the probability of measuring 0 or 1. If the incoming photon is linearly polarized with respect to any direction and the angle to the horizontal plane of the PBS is 0° or 90°, then with a probability of $P = 1$ (100%), i.e. always, the measurement result is 0 or 1. If, on the other hand, the angle is 45°, then the objectively random measurement result is 0 or 1 with a probability of $P = 0.5$ (50%) in each case. In these (and all other) cases, the probabilities result are in agreement with Malus's law (see above). All in all, it becomes apparent that a single qubit at the end nodes of a quantum network represents the simplest form of a quantum random number generator that can be used for QKD.

## 2.7   Quantum Teleportation

The idea of teleporting or "beaming" physical objects, preferably people, from A to B is very common in science fiction.

One example is the popular TV series "Star Trek". The original series was a special hit many decades ago. "Scotty, beam us down to the alien planet," would be a typical quote. What follows is a futuristic mechanism that apparently breaks down crewmembers into their atoms and transfers them to the planet's surface, only to reassemble them there. The crew then complete their tasks utterly unharmed before they are beamed back into the spaceship again. This idea seems to be very far removed from any real science. And so, the public was all the more surprised when the term teleportation suddenly sprang up in science papers, which is reflected for example in the fact that popular science books on this subject suddenly appeared on bestseller lists worldwide.

However, the kind of teleportation that can be proven to exist in the world of science is somewhat less dazzling. This teleportation is not about transporting physical objects, and certainly not humans, but rather about safely transferring quantum information from A to B. Quantum teleportation cannot be used to transmit usable information at superluminal velocity, which—in good agreement with Einstein's theory of relativity—remains impossible. Rather, the extraordinary significance of quantum teleportation lies in the fact that it enables specifically prepared quantum states to be transferred intact without being altered by their measurement. It is also possible to teleport completely unknown states—and in particular, entangled states. This results in completely new technological possibilities for the transmission of qubits, but also for the realization of quantum repeaters. Quantum teleportation is therefore a very promising instrument for the quantum internet. It also opens up new possibilities for the processing of qubits in quantum computers.

## 2.7.1 Teleportation of Qubits

The quantum internet makes it possible to transmit qubits from a sender (Alice) to a receiver (Bob). In long-distance quantum communication, Alice and Bob are spatially separated from each other. In quantum computing, on the other hand, typical distances are exceedingly small. Nevertheless, the fundamental principle of quantum teleportation can be applied to both cases. In contrast to classical bits, it is not possible to measure qubits without changing their quantum state. This is a consequence of decoherence effects. The teleported state can therefore no longer be reconstructed on the sender side after it has been transmitted. This is why it is necessary to establish a classical communication channel between sender and receiver (e.g. a conventional IT connection) where the transmission rate is, of course, limited by the speed of light. The transfer of the qubits, on the other hand, takes place instantaneously, at superluminal velocity, via a special quantum channel. For this, Alice and Bob need a maximally entangled quantum state as their resource, which is destroyed during the teleportation process. It is important to note that quantum teleportation occurs even if the state to be transmitted is completely unknown. What's more, that state can also be entangled with other systems. Besides, it is irrelevant in which physical system the initial state and the target state are present.

## Photonic Teleportation

For illustration, let's take a look at an example. Alice wishes to teleport the polarization state of a photon to Bob. For this purpose, both Alice and Bob first receive a photon from a mutually entangled EPR source (Fig. 2.11). Alice then performs a special Bell measurement on her own EPR photon and the state she wants to teleport. As a result of this measurement process, the quantum state of Bob's EPR particle changes instantaneously because of the entanglement. The measurement results are objectively random, which is why the measured state cannot be known. Alice therefore reports her result to Bob via the classical channel. Bob then manipulates his own state. In this way, he is able to restore the original quantum state. According to the no-cloning theorem, it is not possible to duplicate an original quantum state. It

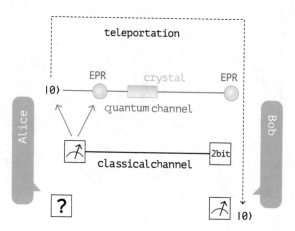

**Fig. 2.11**   Principle of quantum teleportation

must therefore have been destroyed by Alice's measurement. All in all, Alice's quantum information has vanished at location A and was transferred to Bob's location B, which is arbitrarily far away. Again, note that even though the transmission of quantum properties occurs faster than the speed of light, the usable information gain is communicated through the classical channel and thus cannot be gained faster than the speed of light. It should also be noted that the term "transmission" can only be understood in a descriptive sense. Nothing is, in fact, transferred. Rather, a quantum mechanical change of state (determined by the type of measurement) occurs, where the quantum information disappears at one location and is replicated at another location.

### Protocol for Quantum Teleportation

Objective: Alice's quantum information (source qubit) is to be transferred to a physical resource of Bob's (target qubit). The state of the source qubit is destroyed in the process. This is how it's done:

- create quantum channel, create entangled EPR pair
- transmit one qubit of the EPR pair to the transmitter (Alice) and one to the receiver (Bob) via the quantum channel
- Alice performs a special Bell measurement, i.e. a joint measurement of the entangled EPR qubit and the qubit that is to be teleported. This measurement changes the state of the EPR qubit at Bob's qubit instantaneously.
- Alice's measurement result is one of four objectively random states. It is now encoded in two bits and transmitted to Bob via the classical channel.
- Using this classical information, Bob's target qubit can be transformed so that it inevitably assumes the same state Alice's source qubit had at the beginning.

The idea of quantum teleportation has several fathers. It was proposed in 1993 by quantum theorists Asher Peres,

William Wootters, Gilles Brassard, Charles H. Bennett, Richard Josza and Claude Crépeau. In 1997 Anton Zeilinger and his group were able to prove teleportation successfully for the first time, almost at the same time as the group of Sandu Popescu. In all those cases, quantum optical states were transmitted. In 2003, a Swiss team led by Nicolas Gisin was able to demonstrate photonic teleportation over much greater distances. The same team later used Swisscom's commercial fiber-optic networks for another verification. After Innsbruck and US researchers succeeded in teleporting atomic states for the first time in 2004, another Austrian group around Anton Zeilinger and Rupert Ursin "beamed" a state over a distance of 600 m through a fiber-optic line in a sewer below the Danube. In 2012, the same team achieved quantum teleportation in free space between the islands of La Palma and Tenerife over 144 km. The current world record is held by the innovative quantum satellite technology with the teleportation over a distance of more than 1000 km in 2017, which was successfully realized by Jian-Wei Pan and his team.

## 2.7.2 Implementation on Atoms

Figure 2.12 shows (in simplified form) the schematic of qubit teleportation. The qubits are inscribed into individual atomic states. This can be implemented for example with three calcium ions. The timeline can be read from left to right. First, the state to be teleported (for example, $|1\rangle$, $|0\rangle$ or $|1\rangle+|0\rangle$) is inscribed on ion 1 and a special EPR state is prepared on ions 2 and 3. Alice then determines the result of the Bell measurement (states of ions 1 and 2) and transmits it to Bob via a classical channel. The four possible results can be coded in two classical bits, i.e. in the binary numbers 00, 01, 10 and 11. Based on the bit sequence he has received, Bob manipulates his particle

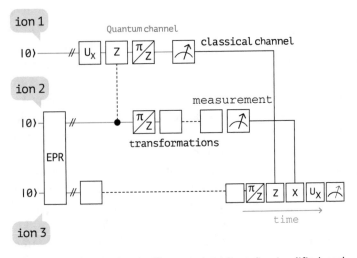

**Fig. 2.12** Teleportation implemented on ions (in simplified and condensed form): The gate operations refer to a qubit representation in the structure of a Bloch sphere. Operation $U_x$ encodes the state of ion 1. The Bell state measurement at ion 2 consists of a controlled Z-gate, a $\pi/2$ rotation and the measurement (a fluorescence photomultiplier in our example). Sequence at ion 3: depending on the Bell base, first rotation by $\pi/2$, followed by reconstruction of Z and X; operation $U_x$ is then used for a fidelity check, followed by measurement via fluorescence multiplying

depending on this information and then performs the final measurement. The state generated by Alice is teleported from ion 1 to ion 3. The no-cloning principle ensures both the inherent security of the transmission and the fact that the quantum state registered by Alice has to disappear if it is replicated by Bob as the original.

## 2.8   Quantum Repeaters

One of the main requirements of communication networks over long distances is a large number of repeater stations. As explained above, in the current internet, data are transmitted as modulated electromagnetic wave trains.

At the individual repeater stations, the signals are measured and amplified and then sent onwards. This technology, which has proven to be very successful in the internet, cannot be used in a quantum network. The fundamental difficulty lies in the fact that quantum information cannot be copied perfectly because of the no-cloning principle. Each measurement would automatically disrupt the qubits to such an extent that they can no longer be reproduced. New technological approaches are therefore of critical importance. A very promising solution lies in the concept of the quantum repeater, which was proposed by the quantum theorists Hans Jürgen Briegel, Juan Ignacio Cirac and Peter Zoller as early as 1998. The basic concept is that rather than amplifying the signal one wishes to transmit, the repeater is used only to build up a certain maximally entangled state. This can then be used in further steps, for example for long-distance QKD with entangled photons. The design or layout has to include a series of quantum repeaters between Alice and Bob, each capable of receiving, processing and transmitting classical and quantum mechanical signals. The construction of maximally entangled states over long distances is achieved according to a special protocol, in accordance with the following three points:

1. Creation of entangled states between adjacent nodes,
2. entanglement swapping (also referred to as entanglement exchange) where the entanglement "spills over" to distant nodes, so to speak, and
3. entanglement distillation (or entanglement purification), a kind of error correction system where a few strongly entangled states are generated from a large number of weakly entangled states.

In practice, steps 2 and 3 have to be performed in alternating sequence. This is because the entanglement exchange requires maximally entangled states. Another essential condition for the design of quantum repeaters is that— even though the losses increase exponentially with the distance—communication should be possible with no more than polynomial increases in the necessary resources (duration, number of stations, number of required qubits, measurements). Currently, numerous theoretical variants and implementation have been proposed for the design, but no decisive breakthrough has been achieved yet. Quantum repeaters have already been successfully demonstrated in various proof-of-concept experiments. Nevertheless, the realization of a technologically viable solution turns out to extremely difficult.

---

### Quantum Repeaters—At a Glance

The main objective of the quantum internet is the entanglement of the qubits at the end nodes. One of the greatest challenges of long-distance quantum communication, however, is that the entanglement of qubits along a "noisy quantum channel" is very difficult. When mobile qubits move via fiber lines to nodes, the absorption and decoherence effects that occur in practice are considerable. Since these losses increase exponentially with the length of the channel, the entanglement cannot be kept intact after a certain distance. To overcome this cardinal problem, the quantum repeater relies on entanglement swapping, where entangled particle pairs are first generated over short distances and then successively entangled by further subsystems to extend over larger distances. Noisy states with minimum entanglement are distilled to purified states with maximum entanglement. For QKD applications, it is also possible to use quantum satellites, which support quantum channels over longer distances. The vacuum in outer space provides a significant advantage for such quantum satellites. The losses over longer distances, including those caused by atmospheric diffraction, are considerably less. All in all, the development of quantum repeaters for a quantum internet requires very complex quantum technology.

## 2.8.1 How It Works

In order to be able to understand the fundamental principle of the quantum repeater, let's take a look at a simplified example. Suppose Alice and Bob perform quantum cryptography with entangled photons according to the Ekert protocol (Sect. 2.6.2). Now, the two are so far away from each other that "path losses" play a considerable role. For example, absorption in the fiber-optic system or decoherence effects may cause the photon rate to be too low for their experiment. What are they going to do? They have to keep the channel as free from noise as possible over a greater distance. In order to achieve that, they use quantum repeaters. In mathematical terms, the method can be represented as follows:

$$if A = B \wedge C = D, \ then A = D$$

where "$=$" corresponds to the entanglement and "$\wedge$" stands for a special Bell condition measurement. Between Alice and Bob there is a repeater station, which first creates entanglement with both Alice and Bob. This process results in two entangled subsystems. As soon as the Bell measurement is performed, the entanglement "swaps" from the subsystems to Alice and Bob. In this way, a pure entangled state is created between Alice and Bob that had not existed before. From that point onwards, it is possible in principle to perform quantum cryptography according to the Ekert protocol. For this to function in practice, however, the qubits have to be stored in a local quantum memory. Current implementation experiments work with photons as mobile qubits, which enter the repeater from both sides via a fiber-optics line. There, the quantum state is stored in a separate quantum memory (for example a trapped atom). By means of a special coupling (corresponding to a Bell state measurement), two stationary network nodes can be interconnected in this way. Similar to the delicate technology of quantum interfaces, the search for robust quantum memories that can interact efficiently with photons proves to be a considerable challenge. Finally, the entanglement must be established faster than the stored states expire.

## 2.8.2  Entanglement Swapping

In the following, the process of "entanglement swapping" is examined in more detail. As outlined above, quantum teleportation enables the transmission of quantum states that are determined concretely, i.e. specially prepared qubits. It is, however, also possible to teleport completely unknown states, particularly entangled states. Quantum repeater technology takes advantage of exactly this circumstance. In a first step, the two entangled pairs $A = B$ and $C = D$ are created (Fig. 2.13). The Bell measurement then causes B and C to become entangled, thereby fulfilling the precondition for quantum teleportation. As a result, the quantum state that corresponds to the original entanglement of $A = B$ is teleported to the $A = D$ system. And so, photons A and D are instantly entangled with each other. It is remarkable that the photons A and D now are strongly correlated, even though they had not been related in any way in the past. This process can theoretically be repeated with any number of repeaters, in this way extending over large distances. Note that entanglement swapping is actually the teleportation of entanglement. Along a series of repeater stations, multiple teleportation processes are performed. In this way, a quantum channel is created. This technology makes it (theoretically) possible to link two quantum computers over thousands of kilometers, which would be very useful for a future quantum internet. At the same time, each station also provides the possibility

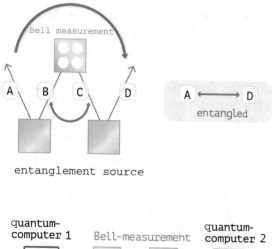

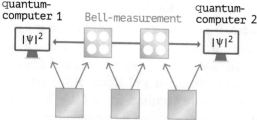

**Fig. 2.13** Schematics of the quantum repeater

for entanglement purification ("error correction"), which makes it possible to maintain a maximally entangled state even over very long distances. Since the teleportation of qubits is only possible with maximally entangled states, this technical design plays a very significant role. Welcome to quantum communication in the world of the future!

## 2.9 Vision and Reality

In view of the enormous difficulties and the fact that an almost completely new technology still needs to be developed, the realization of a (global) quantum internet in the immediate future remains a vision. It is, however, necessary to distinguish between different levels. In a first phase, the installation of larger QKD topologies at various points on Earth and the implementation of several quantum satellites or even drones is quite likely to happen within the next 10–15 years. Leading researchers even assume that within that time period, a quantum internet will already be implemented parallel to the existing internet. First and foremost, it is necessary to establish critical infrastructures on the ground as well as in space, especially now that government institutions and companies are starting to show an interest. To this end, it is necessary to develop systems that can above all be operated economically. Of equal importance is the research goal of developing a network of scalable quantum nodes as prototypes for a fully entangled global quantum internet. This forms an important foundation for the practical realization of quantum repeaters. Also, the software required for scalable control and the networking stack needs to be programmed. It is important to understand that the development of this complex quantum technology must be financed today, for it to be available in mature form tomorrow. As countless implementation attempts show, quantum IT is still predominantly in its infancy. This is true above all for the quantum computer, the development of which probably poses a significantly greater challenge than mere network technology (and should be seen separately from it in principle). Nevertheless, there are numerous efforts everywhere to lift the "Q-web" from the stage of pure basic research to the level of technological or commercial usability.

The European Union, for example, which declared quantum technologies to be the third of its billion-dollar flagship projects in 2016, confirms how seriously this issue is now being taken. The initiative not only aims to harness the benefits for science, industry and society, but also to establish the EU as a global player in this promising new domain. In the sense of a strategic investment, it is hoped to turn Europe into an attractive, dynamic region for innovative business and to create a new level of cooperation between science and industry in order to speed up development. Of course, such plans must not remain lip service. They require appropriate support measures and investments, which should be provided in a timely manner, and in sufficient quantities. While Asia occupies the leading position especially in Q satellite technology, Europe is well positioned in basic research and able to render the existing fiber-optic structure of the internet "Q-fit". Europe should seize this opportunity, pool its resources and shape this international development, not least because many of the ideas implemented in Asia today due to the much larger resources were developed 10–20 years ago in Europe and the USA.

The EU Commission has asked leading representatives from science and industry to present their ideas for a realistic roadmap with targets and timelines. This shows that the quantum internet is embedded in a 4-column model that comprises the core regions characterizing the potential second quantum revolution: Quantum communication, simulation, sensor technology and computing. The EU Commission recently approved a first grant for the prototype development of a scalable quantum network. The grant was awarded to the "Quantum Internet Alliance" (headed by Qu Tech University Delft), a Europe-wide consortium of leading research institutions. The objective of this association is the development of the necessary technology, in collaboration with partners from industry and high-tech companies. Of course, to position members at the forefront of this highly innovative domain is in the interest of this coalition. Of special note in this context is the British Quantum Communications Hub, a research and development consortium of universities, industrial partners and governmental interest groups funded by the British National Quantum Technologies Programme. Its focus lies on commercialization. The development of a powerful quantum computer is envisioned as an even more distant, but all the more fascinating, objective, which is to grant Europe's future smart industry vision the ultimate triumph. With such bold ambitions, Europe is in the good company of countries such as Japan, China or the USA— seen by Europe not only as competitors, but also as possible collaboration partners. In particular the considerable interest of American tech giants in quantum computers, but also the fact that quantum information technology is among the most important commodities of the future, makes the vision of a quantum internet appear in a much more realistic light than some would like to admit.

Assuming scientific and technological feasibility (which has not yet been conclusively proven), three key areas emerge for the quantum internet (see box).

### Development Areas of Quantum Information Technology

- Establishment of local, later also global QKD networks as a key element of long-term digital security technology. Usability of QKD for numerous commercial applications and mobile devices. In addition to satellite-to-ground data links, one of the declared research goals is to change the fiber-optic structure of today's internet in such a way that maximally secure data transmission covering all five continents is achieved. The technological objective describes a network system of quantum nodes for fully entangled QKD without the need for trusted repeaters. The use of quantum memories is required for repeater technology, but not for the users at the end nodes.
- Development of powerful quantum processors, ideally scalable quantum computers, whose processing power is available via quantum cloud to users from science, medicine and business, but also to private users.

The associated possibility of blind quantum computing necessitates a network of reliable quantum memories where users can (ideally) prepare states, store qubits and teleport. The basic functionality was demonstrated in laboratory tests. Its practical realization, however, remains a distant goal and requires a number of technological breakthroughs.

- Global networking of scalable quantum computers and quantum devices of various types and designs. The end nodes are powerful standalone quantum computers with the ability to make reliable error corrections. They also enable the generation and teleportation of complex quantum states. If this stage of development is actually reached one day, this would result in a kind of universal quantum internet for multi-user applications with probably still completely unknown applications and possibilities for expansion.

## 2.9.1 Agenda 2030—The First Global Quantum Internet?

QKD is the technologically farthest advanced area out of all quantum technologies. As outlined above, true network testbed environments have been implemented in Europe, the USA and Asia. Japan, but also China are the pioneers in this area. For the Beijing-Shanghai backbone and the QUESS technology demonstrated there, the obvious objective is a global quantum network. In a next step, the existing terrestrial network will be expanded considerably and operated by a start-up company founded especially for that purpose. This pilot project will serve as an example for possible local networks, which could in the long run be connected via quantum satellites on a globally mobile basis. As a target period, the years around 2030 are aimed for. Chinese chief scientist Jian-Wei Pan estimates that global coverage could be achieved around that time via several Q-satellites. What might that look like in practice? One possibility was demonstrated recently in a quantum telephone conference between China and Austria. However, one drawback is that for very long distances, the satellite can only serve as a trusted repeater. This implies that trust in the satellite operator is necessary. Nevertheless, the developed "relay technology" already represents considerable progress and proves the technological feasibility of long-distance quantum communication. It should also be noted that the (maximally secure) realization of fully entangled QKD over 1200 km would already be possible. That's enormous progress compared to previous implementations! It is therefore highly likely that China will further develop the "Q-Sat technology" into a Q satellite network from an economic point of view. Any progress in classical optical satellite communication is likely to provide significant support in this respect. With appropriate logistics, even several quantum satellites may theoretically be interconnected briefly to form a quantum repeater, whereby very distant end nodes are directly entangled by means of entanglement swapping. Provided that generation rates are sufficiently high, key quantum material might be generated in this way which, after several cycles at the latest, will reach orders of magnitude that enable highly secure transmission. As shown above, with a fully entangled QKD there is no physical possibility to pick up the generated key from the satellite and transmit it to third parties. Leaving aside the technical difficulties, this would not only solve the repeater problem (with some limitations), but also create quantum channels between local networks, which by then might be in existence all over the world. According to the estimation of some researchers, we can expect considerable further developments in quantum repeater technology in the next 10–15 years. This means that by then, the fully entangled

QKD will also be possible in fiber-optic networks over long distances. From that point onwards, the existing fiber-optic infrastructure of the "old" internet can also be used globally for quantum communication. This is particularly important for European research, as the Vienna Multiplex Q-Web proves. Under the headline "Vision Web Q.0", the QIA (Quantum Internet Alliance) is developing a prototype quantum internet consisting of 3–4 nodes, which will connect four cities in the Netherlands using quantum repeaters. This is not only to demonstrate how quantum repeaters work, but also to enable the first quantum network that allows the exchange of qubits between small quantum computers. Efforts are underway to realize a first test link as early as 2020, which will demonstrate the feasibility of adapting existing optical fibers via frequency conversion methods, among other things. If successful, the project would be an important milestone on the way to the development of a "real" quantum internet without trusted repeaters. In the following, the two most important functionalities of the QKD are presented.

## 1. Resistance to Supercomputers

One advantage of QKD that should not be underestimated is its ability to render the swift yet vulnerable symmetric encryption secure. As mentioned above, classical key allocation would remain a serious problem if done via one-time pad (OTP) encryption (Sect. 2.6.1). Still, OTP is a useful method not only for the encryption of very large amounts of data within very short time frames (which will play a major role in future IT), it is also able to withstand attacks from future supercomputers and quantum computers at sufficiently high bit rates. From today's perspective, the latter only has a chance against OTP by using a kind of Grover algorithm. As shown, even though the Grover algorithm is able to achieve "quadratic acceleration", this is not sufficient against OTP in view of the myriads of possible key combinations. To succeed, the quantum computer would have to come up with a surreal qubit number (note redundancy!), which is completely unrealistic from today's perspective. As was shown in detail, with QKD, security based on the laws of physics takes effect. Because of this, any eavesdropping attack can be detected directly and without delay. This functionality is not even possible in classical IT, it is innovation in its truest sense. This feature will make QKD interesting for private users and commercial applications beyond the interests of governments, secret services, the military or large corporations. Relevant products and services in the field of finance management include digital payments or very secure ATMs and credit cards. The psychological factor of trust may also play a role. It is simply more reassuring to depend upon security guaranteed by natural laws than on the repeatedly proven fallibility of human beings. The motto "security built by nature" comes to mind. However, numerous technological hurdles still need to be overcome before quantum channels over long distances can be realized, including the repeater problem. Such advanced systems also form the basis for the visions of classical IT. A network of future smart cities for example would hardly be conceivable without special safety technology. QKD offers an ideal technological foundation for this, which, together with methods of post-quantum cryptography should also be able to withstand future attacks with dramatically increased computer performance. What QKD is not able to do, however, is to provide complete protection against fake authentication. But even this problem, which from today's point of view is generally unsolvable, can be considerably alleviated by QKD-based procedures.

## 2. Hacker-Proof Data Storage

Although QKD per se only permits entirely tap-proof point-to-point data transmission, it can also guarantee unequalled levels of security in combination with classical methods. One example has already been implemented in test networks. Even today, enormous amounts of data are stored in data centers over very long periods of time. Also, digital archives are increasing in size. How can comprehensive protection possible against hacker attacks be created, even if those attacks make use of future computer performance levels? A future-proof data storage system therefore has to meet the following four requirements:

1. Confidentiality (data is only accessible to authorized parties),
2. integrity (data remains unchanged, i.e. use of digital signatures and authentication schemes),
3. availability (data can be retrieved at any time through redundancy) and
4. functionality (data can be processed further without decryption, which requires so-called homomorphic encryption).

A possible approach is presented in the following. From a central set of data, parts are stored on various distributed memories using polynomial multiplication. With an amount of $N$ memories, the data can be reconstructed by collecting at least $k$ data packets. If the number of packets is $k - 1$, the data cannot even be reconstructed if the computing power is unlimited (provided that the number of corrupt partial memories is less than $k$). This system guarantees confidentiality (1) while it is possible to algorithmically multiply its parts so that (4) is met. Even if parts are lost, the data can be reconstructed, fulfilling condition (3).

Requirement (2), on the other hand, is not necessarily guaranteed. Above all, the communication between the partial memories has to be protected, which is solved ideally using QKD. QKD therefore enables long-term integrity and confidentiality protection for hacker-proof data storage devices.

## 2.9.2 Future Zone: The Universal Q-Hypernet

As the number of qubits at the end nodes increases, our quantum network gets more efficient and powerful. Given that the repeater problem can be resolved and that coherence states as well as error correction systems can be optimized continuously, one day, a global network may well emerge which, as an extremely powerful hypernet with additional quantum power, would complement visions in classical IT. Based on extremely complex network technology that can be coordinated with ultrafast speeds and expanded using numerous quantum satellites (which theoretically can be reduced in size to microformats), it would be suited to take humankind into a completely new technological era. Such a hypernet could also contribute to numerous innovations outside science (where its value would be inestimable). These include absolutely secure QKD-protected communication channels (in their ultimate form on a fully interconnected basis) which connect classical computers with varying mobile terminals into networks. Examples for such end-user devices include smartphones, wearables, satellites, drones and self-driving cars. Such technologies would be especially effective as protection against hacking attempts attacking databases and digital archives, which will become even more prevalent in the future, providing a multitude of commercial applications for the average user.

Besides highly secure online financial transactions, which would already be essential then, other applications are conceivable as well. Thanks to the superposition principle, quantum blockchains provide the basis for completely new methods of validating offers, contracts or crypto currencies such as bitcoins. Similarly, it would be much easier to efficiently identify fake news. Another aspect supports the increasing need for privacy, which is likely to become more important in the future. At the moment, people (especially young people) are basically entrusting their entire lives to the internet in the form of digital footsteps. In this way, they throw themselves at the mercy of algorithms and institutions. Several internet services contribute considerably to this tendency. Even today it is difficult to keep private characteristics and preferences hidden. With every entry into the search engine, personal wishes and desires are recorded for posterity, or for advertising companies. Quantum technology provides (if used accordingly) the opportunity to defend an important social value: the right to privacy and protection from becoming completely "transparent" and manipulable. A super-safe quantum cloud would also fulfill this need.

A technological highlight is the development of scalable quantum computers, the advantages of which will be available to users all over the world. The "quantum advantage" could one day be immense for economy. Not only cloud operators will make fortunes (remaining unaware exactly what functionalities their clients perform). Consider all the different aspects of the necessary infrastructure! Quantum technologies might one day be an Eldorado for all kinds of new companies. Completely new professions and business ideas will likely emerge. In the quantum internet of the future, the user enters a question into the quantum search engine and receives an answer without anyone knowing the question—not even the server! However, since search engine operators today earn their money by analyzing user data, they will inevitably have to come up with other sources of income. Customers might be offered the alternative of whether to disclose the contents of their search query or to pay extra for a highly discrete search. Besides, business travelers will be able to have their optimum route calculated. Pharmaceutical companies will get the necessary computations for their new miracle drug. The implications extend to biochemistry and genetics and beyond. Countless companies will have their logistics worked out, and traffic authorities will be able to rely on perfect calculations of vehicle flows. Quantum computing also can be applied to financial problems by identifying complex variable relations, e.g. the classification and selection of assets, customers, or vendors by default risks. Thanks to quantum simulators, science will develop completely new materials or realize superconductivity at room temperature, which would take entire industry branches into a golden age. Companies will be able to develop more efficient concepts for batteries and their recycling and in this way also optimize e-mobility. As today's studies already suggest, quantum computers open up undreamt-of possibilities for machine learning, robotics and AI, which would really get smart industry concepts off the ground.

After all, powerful quantum computers and devices of any kind may one day be linked by pure entanglement or exchange complex quantum information with one another using quantum teleportation. True to the Feynman doctrine that quantum computers represent nature, further future technologies such as "programmable matter" can be derived from it. Theoretically, some kind of quantum 3D printer would be conceivable, which would download complex quantum information from the quantum cloud via teleportation onto an existing material object carrier.

Such technology would enable us to design the properties of matter on a microscopic scale and adapt them according to the client's wishes. Another possibility is that all this will one day lead to intelligent XD materials that adjust adaptively to their environment. At the start of this book, the networking of quantum computers via teleportation was mentioned as the underlying principle. This vision is guided, among other things, by the idea that the entire quantum network might become an extremely powerful modular mainframe computer in the sense of networked computing. Because the repeater problem can basically be solved on the basis of multiple teleportation, this decentralized concept may play a key role in future quantum IT. The realization of powerful quantum computers will probably represent one of humanity's greatest technological achievements in the coming decades (and far beyond). In any case, its development is closely linked with the quantum internet. Besides blind quantum computing, such a system—on whatever scale—is necessary in order to develop a scalable quantum computer at all. The quantum internet might also support networked quantum computers to reach even higher performance capabilities. From a fundamental physics perspective, such futuristic visions are conceivable in principle. The question is to what degree quantum coherence can be maintained on a large scale, in order to technologically even come close to the possibilities that nature offers in principle. Quantum information technology has tremendous potential. If the "second quantum revolution" actually comes to pass, the universal quantum hypernet will appear as its crowning ornate.

# 3

# For Deeper Understanding

## 3.1 Workshop: Quantum Optical Systems

**Quantum Interference**

Quantum optical devices play a major role in quantum communication technology, particularly for QKD networks. To better understand their significance, the reader is invited to a fictional Institute of Experimental Physics, where "Professor Quant" will answer questions. He will explain what we mean by wave-particle-duality, how the wave function functions is or what nonlocality is. Such terms are likely to cloak the quantum world in mystery, probably due to the fact that our human imagination is so tenaciously rooted in the world of everyday experience.

For starters, the professor shows us a Mach-Zehnder interferometer (Fig. 3.1). Such devices are used in technology, for example to detect minute density differences

© Springer Nature Switzerland AG 2020
G. Fürnkranz, *The Quantum Internet*,
https://doi.org/10.1007/978-3-030-42664-4_3

in materials. In modified form they serve as inertial sensors in airplanes. They are also important for experiments to prove fundamental principles in quantum optics. Note: Fig. 3.1 shows two experimental setups, one using screens and an ordinary laser, the other one using detectors and a single photon laser.

First, Professor Quant explains the arrangement using screens and conventional lasers. In the setup, we can see how a laser beam hits beam splitter 1, where it is separated into two beams. Each of the resulting beams is deflected by a 90° mirror. Then, the two beams merge again in beam splitter 2. When Professor Quant flips the switch on the laser to turn it on, a peculiar pattern of stripes appears, the so-called interference fringes (Fig. 3.1 top). We see the image emerge both on screen 1 and screen 2, which are positioned exactly where the beams exit the setup. A diverging lens (not shown) is necessary to make the pattern clearly visible to our human eyes. Professor Quant sticks his hand into one of the partial beams and shouts excitedly. "You see?" We see nothing except that the pattern disappears at the moment he holds out his hand. "Do you understand now?" He asks emphatically, persistently. But we do not understand. What's that supposed to prove, please?

Patiently, the professor explains what this is all about. "You see, from a very modern point of view in quantum physics, you have just observed a 1-bit system. You know that light consists of photons, of light quanta. Whenever a photon is sent out by the laser, it has a 50% chance of being transmitted or reflected quantum randomly at beam splitter 1. As long as I do not place my hand into its path, it is impossible to determine which path the photon has taken. We say that the system then is in a superposition. That means, both possible partial beams, and the two paths the photon can take, are superimposed on

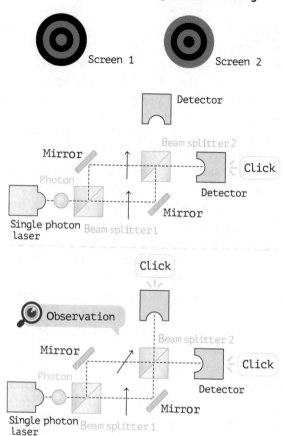

**Fig. 3.1** Well-adjusted Mach-Zehnder interferometer. The single photon laser generates single light quanta, which are then measured via the beam splitter arrangement of detectors. Depending on the position of the polarizing filters (black arrows), complementarity between path information and interference is demonstrated. *Note* Fig. 3.1 indicates a second experimental setup where the detectors are replaced by screens 1 and 2 using a conventional laser. This is mentioned in the text, to help readers reach a better overall understanding

each other. When this happens, the interference pattern appears. Now, the instant I stick my hand into one of the two partial beams, it is immediately clear which path the photon has taken, since it collides, so to speak, with my hand when it has chosen this partial beam. But if I don't, then we cannot know which path the photon has taken. It's important to note that the collision does not have to happen from the point of view of a single photon, it's enough that it could have happened! We then speak of *location measurement*. At the moment of this measurement, the superposition collapses immediately, and the striped pattern disappears. Niels Bohr, one of the founding fathers of quantum mechanics, referred to this as the collapse of the wave function. Surprisingly, the behavior of the system does not depend on whether the photons are observed or not, but on whether they *could* be observed or not."

"So, you're seriously saying, as a scientist, that quantum systems depend on observation?"

"In a way, we could say that. Now, to 'observe' does not automatically mean that we really have to detect the particle. The mere possibility is enough. To observe here means to, in terms of physics, to create a situation where it is possible to measure one or more physical quantities. I don't necessarily have to hold out my hand for this. Typically, we would take two polarizing filters and insert each of them into one of the two partial beams. Polarizing filters have the property that they change the vibrational plane of light. Here, I will show you. You see? As long as both polarizing filters (black arrows in Fig. 3.1 top) are parallel to each other, the interference image remains in its familiar form. However, if I rotate the filters against each other (Fig. 3.1 bottom), the interference disappears. The reason is that a position measurement occurs in that moment, since the twisted vibrational planes of the light could be used to determine the partial beam in which a certain photon arrives. So, there is a certain amount of path information. And if I now set the filters parallel again, this info option disappears, and the system again creates the interference pattern."

"And what does that have to do with a 1-bit system?"

"Quite simply, only two mutually exclusive possibilities exist. It's either interference *or* position measurement. Neither is possible at the same time. For example, 1 bit of information can mean yes or no, white or black, off or on, cold or warm, and so on. The either-or principle always applies. In the language of quantum physics: interference and path information are complementary to each other. If we continuously rotate the filters from their parallel position, we see that the interference decreases successively until it disappears completely at 90°. So, at the expense of interference, I am receiving more and more path information. Conversely, if I slowly rotate the filters back until they are again parallel to each other, the path information decreases continuously in favor of interference. This is exactly what complementarity means. The more knowledge (i.e. information) we have about a certain value or property, the less information I have about the other one. In total the system contains no more than 1 bit of information. Niels Bohr understood that complementarity is a fundamental principle of quantum mechanics and applies to all quantities or properties that can be measured. It not only serves as a limit to what we can know, but also to what properties a system can possess at all" (Zeilinger 2005a).

"Now let's be honest. This entire thing seems somewhat artificially construed. These phenomena can easily be explained by classical physics. Of course the second partial beam is missing if your hand obstructs it, and this excludes interference per se. The polarizing filters are also easy to explain. The interference images that occur require coherent partial beams. But if both partial beams are not parallel with respect to their plane of oscillation, this condition is automatically not fulfilled. So what do we need quantum physics for at all?"

"Excellent!" Professor Quant praises us. "You clearly know something about physics—well, at least about classical physics. In fact, I agree with you! As long as our laser beam is visible to the human eye, for which it needs to carry millions and millions of photons, quantum theory is not really necessary. If, however, the beam is attenuated to such an extent that it is made up of individual photons only, that's a different story. A little patience, please. Things will become very, very interesting!"

"And by the way, what do you mean by the term 'wave function'?"

"Exactly!" Professor Quant laughs out loud. "The wave function is a curious thing. Think logically! When the photons hit the first beam splitter, they have a random quantum probability of 50% to be either transmitted or reflected. A single photon in the "lower" partial beam has a 50% chance of being transmitted or reflected at the second beam splitter. Statistically this means that 25% of all photons are transmitted at beam splitter 2 and 25% are reflected at this same beam splitter 2. If, however, the photon has chosen the "upper" partial beam, this splitting also occurs at beam splitter 2, i.e. 25% are transmitted there and 25% are reflected there. In total, half of all photons should reach screen 1, the other half screen 2. What we should see then would be a diffuse, unstructured light spot on each of the two observation screens. But that is not what we actually see. Instead, we see the pattern of stripes on both screens, the interference fringes. Even though here, too, the number of photons is 50% for each screen. Still, we have no explanation at all for the occurrence of this sequence of darkness and light.

"What's so particularly surprising about that?"

"What's so astonishing is this. We are in a veritable dilemma, because the experiment shows a different result than the theory we just established would predict. Experiment, however, is the highest judge in physics, and so the theory has to be changed so that it corresponds to the experimental findings. This is done by describing light as a wave. But we have to be careful! The experiment also creates another problem. Experiments in accelerator facilities (such as CERN) have confirmed that light definitely consists of particle-like photons. One possible way out would be to say that light, without exception, is made up of photons—but how these particles are arranged on the measuring screen can only be described consistently using the model of a wave. In principle, however, this wave image is a fiction, a mathematical tool that makes it easier for people to think about these strange phenomena. For fundamental reasons, it is never possible—in quantum mechanics—to predict exactly where a certain photon will hit the observation screen. All that is possible is to calculate a probability. Therefore the wave function, or more precisely its amplitude square, has the character of a probability wave. Imagine it like this. Light consists of particles. The probability of measuring these particles at a given location is determined by the amplitude square of the wave function.

"Then how does this probability wave model explain the phenomenon?"

"It's very easy to explain the emergence of the interference fringes on the basis of the wave concept. The explanation will naturally correspond to classical physics. We now assume that the two polarizing filters are parallel to each other. At the first beam splitter, the laser light is split into two partial waves, each with half the original intensity. If you follow the upper partial beam in Fig. 3.1, you see that it is reflected twice before hitting screen 1. If the lower partial beam is also to hit screen 1, it will be reflected two times as well. Both beams are reflected by a beam splitter and a mirror. Since the two beams are reflected in the same way, they have a so-called phase difference of 0, which corresponds to a constructive interference. This means that the amplitudes of both partial waves add up, creating a wave that in sum doubles their amplitudes. Because the amplitude in the wave image corresponds to the brightness of the light, all of the brightness and accordingly all of the light from the laser is now visible on screen 1. None of the light remains for screen 2. In what is referred to as a well-adjusted interferometer (see below), screen 1 would be very bright and screen 2 would be completely dark.

Now, the well-adjusted interferometer is an idealization. Such abstract representations are common in physics. They are tools to enable mental work. In the case of the well-adjusted interferometer, it is assumed that all optical path lengths are completely identical. In practice, however, this is difficult to achieve because the laser beam, despite its slim shape, always exhibits a certain divergence. Therefore, in our simplified experiment here, the optical path lengths are usually not exactly the same. The result is a sequence of constructive and destructive interference from the inside out. Accordingly, an interference pattern of light and dark rings is observed on both screens. However, the two interference images are complementary, i.e. if there is a bright spot at a certain point on screen 1, darkness prevails at the same point on screen 2 and vice versa. Exactly this circumstance is described with great precision by the wave model. In particular, it is the wave model that makes it possible to understand how light— even just a single photon!—is able to obliterate or amplify light. Without the wave image of light, physics would not be able to provide an explanation for the above experiment. That's why it is necessary to attribute wave character to light in the mathematical sense."

"What happens if we replace the screens with two photon detectors, as is done in quantum cryptography?"

"That's a very good question. I was about to demonstrate exactly that just now. We switch to high-tech experimental physics mode. Let's take a well-adjusted interferometer and use two photon detectors instead of the two screens. The detectors are able to detect individual photons using a sophisticated multiplication technique based on special photodiodes. Our key modification is to replace our laser with a single photon laser. Let's see what happens next. Statistically speaking, there is now only one single photon inside the interferometer. Of course, our human eyes are not able to see this, because the laser light is much too weak now. Our super-precise detectors, however, are able to register this single photon. You see? If the polarizing filters are rotated 90° to each other, both detectors respond, that is, in sequence, one after the other. If, however, the polarizing filters are parallel to each other, only one of the two detectors "clicks" at a time. This is due to the complementary interference pattern I explained before. Note that the well-adjusted interferometer no longer produces a pattern. Rather, the entire intensity of the light—that is, the number of photons—arrives at one single screen or detector. And now I am asking you this. How is it possible that individual particles exhibit such strange behavior? If both detectors respond, this can still be described with the pure particle concept, as we considered above. The second case, however, where the polarizing filters are parallel too each other, that's a different story. There's no other way than the wave model to explain this case. You see, depending on the experimental setup, the light exhibits either particle-like or wave-like behavior.

This strange property of quantum objects is often referred to as the wave-particle duality. We need the wave function to describe this ambivalence."

"Does the wave function refer only to light quanta or is this a universal description of quantum physics?"

"It's definitely a universal description. It can be much more complicated mathematically than in this simple case. But it pops up again and again in quantum physics. Sure, yes, the wave function, or the $\psi$-function, is an absolute necessity. It is not just photons that behave in such a strange way. All particles, i.e. quantum objects, are subject to it, including electrons, protons, neutrons, entire atoms or even rather large molecules. We'll get to that later. But first, I would like to show you another very famous experiment. Imagine a wall with two tiny openings, just big enough for a quantum particle to pass through. This wall receives a barrage of many, many projectiles, like from a machine gun. These 'bullets' aren't necessarily massless particles, such as photons. The ammunition can also consist of particles with a rest mass, for example negatively charged electrons".

"Are you now talking about the famous double-slit experiment with electrons?"

"Yes, that's right. An enormous collective of individual electrons is emitted statistically randomly from a source. They smack into a wall with two tiny openings, i.e. a 'double-slit'. Many electrons are absorbed by the wall, but a few make it through the two slits. Each electron has a 50% chance of passing through either one or the other slit. On average, therefore, the same number of electrons will pass through each opening. The particles are registered on an observation screen behind the wall. One would expect a frequency distribution that looks a bit like kicking a large number of dirty soccer balls through two openings onto a white wall. Each ball leaves a stain on the wall behind it. So much for the theory. However, if the experiment is carried out in a real-world setting, the frequency distribution that actually appears is a very different one. I cannot demonstrate the setup here; the experiment is too complex for that. It was first performed in 1957 by Claus Jönsson and then even demonstrated with entire atoms by Jürgen Mlynek and Olivier Carnal in 1990, each time with the same result. The image that appears always has a certain similarity with the interference fringe pattern from our previous experiment. Indeed, it's an interference pattern again. It's very strange. The quantum interference that occurs in the electron experiment corresponds to a kind of superposition of the two options that an electron has—that is, to pass through either one or the other slit. Weirdly enough, the two point-like electrons behave as if they had passed through both slits at the same time—which, of course, cannot be the case. So in a mathematical sense, a wave aspect must be assigned to them. On this basis, it is possible to calculate the probability with which the electrons hit the screen. This can be understood in the sense that although each individual electron behaves as a point-like particle when registered on the screen, the overall distribution is determined by wave-like probabilities. From the previous experiment you know that interference images can in general only be explained on the basis of the wave concept. As you can see, the concept of the $\psi$ wave function is therefore not only limited to photons but also affects material particles. In this case we speak of matter waves. Generally, all quantum particles exhibit this remarkable wave-particle duality.

By the way, let's see if you are already able to think in quantum mechanics terms. In the double-slit experiment, what will happen if we close one of the two slits? That's right. This corresponds to a position measurement, since at this exact moment, the electron reveals through which of the two slits it has passed. The observer there-fore receives location information about the electron. The result should be that the interference image disap-pears immediately—which it does! Or more precisely, the interference pattern is significantly weaker than before. The actually detected image looks like the pattern many dirty footballs leave when kicked through a single slit. Et voilà! Here, too, we have the principle of complementa-rity. It's either an interference *or* a position measurement. They cannot both be realized at the same time. Thus, the double-slit experiment behaves completely analogous to our Mach-Zehnder interferometer and shows the same 'information-related' behavior."

"The quantum world really seems to be pretty crazy! Particles that are measured as particles but otherwise behave like waves so that they can somehow pass through slits at the same time? That can't be serious science! Isn't it possible that there is some kind of interaction process behind this? After all, visible light is made up of tril-lions and trillions of photons. Maybe they just obstruct each other, or bounce off each other somehow? The par-ticles might just as well divide, couldn't they? Then one could simply forget about the contradictory wave-particle aspect?"

"Of course, that seems like a sensible suggestion. But you see, you can also do the double-slit experiment with *single* electrons and it produces *the exact same* results. So, you can shoot many single electrons one after the other through both slits, and then register each single point of impact. To your surprise, you will discover that a wave-like interference pattern is created again. Besides, you just saw the Mach-Zehnder interferometer with the single photon laser. There, at any given time there is only one single photon in the interferometer and *still* there is interference! Any kind of hidden interaction is inevitably excluded. What's more, the whole thing borders on magic. Remember that whether interference occurs or not depends on the position of the polarizing filters in relation to each other. But how can a single photon know which setting the two polarizing filters currently have? How can it know that, if it can quantum randomly take only one of the two partial beams at a time? According to our human perspective, it would have to divide itself somehow at the first beam splitter in order to be in both beam paths at the same time. But as we know, that is impossible. Photons are indivisible. We can easily demonstrate this by removing a beam splitter from the Mach-Zehnder and directing the single photon source at the remaining one. If we now place the photon detectors at the two exits, we see that alternately only one of the two detectors "clicks". Never both at the same time. The photon is not divided. It is either transmitted or reflected quantum randomly with 50% probability. This is a quantum random number generator, as it is also used in QKD. By the way, it is also impossible to divide the quantum random coincidence further. It is the most elementary component of a quantum mechanical event. And that's why the random numbers generated by QKD are of the best quality that's achievable.

So, now you can see that quantum physics puts human imagination to the test. This property of quantum particles, namely that they "sniff out" their surroundings, as Richard Feynman put it, is commonly referred to as nonlocality. To describe something like this, we need the wave function. Its actual significance is still a very controversial issue in physics today. But we can also forget about the deeper meaning and work with the wave function as an abstract auxiliary construction that helps people to understand the rules of quantum physics. You can see that the concept of information plays a very significant role here. Depending on the experimental setup, location information may or may not be present. This depends on the position of the polarizing filters in relation to each other (as with the Mach-Zehnder) or whether both slits are open, or one is closed (as with the double-slit experiment). Accordingly, information has a direct influence on the behavior of the particles, so to speak. Even if we want to speak about what happens in a physical experiment as "reality", it is in any case directly related to the concept of information. According to Anton Zeilinger, information even is the most fundamental building block of the universe" (Zeilinger 2005b).

"Earlier, you mentioned that the location where a single particle hits the screen is subject to statistical randomness. Why can't we predict that exactly?"

"Well, philosophers are still arguing about that, too. I boldly depart from the usual statements and explain it like this: Because for some reason, the quantum particle does not contain the full information for that. This is expressed, for example, by the Heisenberg uncertainty principle, which expresses complementarity between place and impulse. I'll show you!"

Professor Quant digs out a small laser pointer and points it at the blackboard. All we see is the typical little round spot of light. He then holds a small, thin platelet in front of the laser and asks, "What can you see now?" We see a pattern of long stripes with a symmetrical sequence of darkness and light. "This," explains the professor, "is another interference picture, of course. The interesting thing is not the interference itself, but how it comes about. The immediate cause is the phenomenon of diffraction. You could compare this to water waves that diverge laterally when they pass through a narrow space or encounter obstacles. Something like this also happens with light. Typically, the diffraction of light is explained by the formation of new waves along a wave front. That effect is the

Huygens-Fresnel principle. I will give you a different, more contemporary explanation: The diffraction of light can also be interpreted as a quantum effect—especially in experiments with single photons.

Inserted into the professor's platelet is a microscopically small slit through which the light has to pass. This gap is so small that a quantum of light, to use graphic language, barely passes through. In terms of information theory, this is a position measurement, because if the slit size is in the range of the photon dimension, we know how large the photon is. Now we have already seen in the previous experiments that the complementary property automatically disappears when a location measurement occurs. Until now, that property has always been interference. Here, however, the situation is reversed. The interference image is created—in contrast to our earlier experiments—as a result of the position measurement. For this reason, this quantum system must logically be another complementary property. For us, a new value is introduced: the impulse of the photon. This is in principle defined as the product of mass and speed on the one hand, yet on the other hand it is connected with the quantum of action and the amplitudes of the light waves or matter waves (I will come back to this in a moment). Photons do not have a rest mass, but they can be assigned a mass equivalent, especially as they transport energy and are equivalent to energy and mass according to Einstein's world-famous equation $E = mc^2$.

So how does Heisenberg's uncertainty relation come into play? Casually speaking, it means that the product of position uncertainty and momentum uncertainty is at least as large as a very, very small number, Planck's quantum of action. Applied to diffraction, this means that the position measurement at the slit reduces the position uncertainty. This in turn means that the momentum uncertainty has to increase. It can also be said that the more position information there is, the less information there is about the particle's momentum. The vector character of the momentum (i.e. its sense of direction) emerges when we see that the laser beam does not pass straight through the platelet. Rather, it disperses. This corresponds to the diffraction, as mentioned. It's only logical that the diffraction angle increases as the slit size and consequently also the position uncertainty decreases. This can also be shown mathematically."

"You have explained the diffraction, but not the interference. How does the interference come about?"

"But you already know that! Based on the wave image, interference is always achieved by superposition of different constructive and destructive components. This is simply a consequence of diffraction—and of the wave-like properties of quantum objects. We were able to show conclusively that these are necessary. However, diffraction is so very interesting because a very important fundamental principle of quantum physics applies: Heisenberg's uncertainty principle, which shows that quantum systems obviously contain a limited amount of information that can be distributed differently. It's either in the location information or in the momentum information of a particle, or in one of the innumerable gradations between the two. The more I know about one of the two values, the less I know about the other—and vice versa. However, it is never possible to obtain the complete information about both complementary quantities at the same time. This is not rooted in our subjective ignorance. Rather, it is one of nature's fundamental principles. A very important conclusion can be drawn from this, which has far-reaching consequences. A quantum particle can never be assigned a clearly defined trajectory, where its exact position is determined by a mathematical function as a function of time."

"This all sounds extremely theoretical. Does the Heisenberg uncertainty relation have any other meaning, apart from your example? Something that's relevant for normal people and non-physicists, too?"

"Khhhh!" Professor Quant splutters. "It seems I have not expressed myself clearly. Probably the most significant effect of the uncertainty principle is that on humans. What's more, we humans would not exist without it. You see, the uncertainty relation is at work not only in our little experiment here, but in every single atom of the entire universe! I just wanted to point out that a quantum particle by nature has no defined trajectory. What do you think, why do they teach us about orbitals, electron clouds or delocalized electrons in chemistry class? All these are effects of the uncertainty relation. Only on its basis is it possible to scientifically explain the formation of atomic bonds and in consequence the emergence of organic molecules, the building blocks of all life as we know it. In fact, our human DNA would have never developed without the uncertainty principle! Even if it conforms to the rules of classical physics as far as genetics are concerned.

Another thing. What would today's world be without smartphones? Unimaginable for the vast majority of people today! Are you aware of the fact that we owe this little piece of everyday technology to quantum physics? Modern microchips are based on semiconductor technology, which in turn is based on solid-state physics, where the lattice structure of atoms and molecules is investigated. And with that, we have again arrived at Heisenberg's doorstep. By the way, the discovery of the semiconductor transistor—the basic component of every single electronic computer—is the achievement of three American quantum physicists. But let's not go into that. I could give you countless further examples where in the end, we arrive at the realization that the entire universe would not exist in the form known to us—without the uncertainty principle."

"If our world is so strongly influenced by quantum physics, why don't we notice its strange laws in everyday life?"

"First of all, this is due to the fact that the Planck constant, the quantum of action, is so extraordinarily small. With about 6.5 ten thousandths of a trillionth of a trillionth of a joule second, this value is so tiny that it remains many powers of ten below any measurement accuracy relevant for macroscopic systems. This leads to such minuscule deviations in everyday objects that they are insignificant for our everyday experience."

"Then where exactly can this boundary be drawn? Where does quantum physics end and where does classical physics as we know it begin?"

"Very good question. Honest answer: We do not know. All we know is that quantum effects get 'washed out' more and more with increasing size, let's say, the larger the mass of the respective object gets. The quantum formalism then automatically merges into the classical formalism. Nevertheless, macroscopic quantum effects that we can see with our own eyes do exist. One example are superconductors. It's a remarkable experience to observe a ceramic superconductor hovering above a permanent magnet. However, where exactly to draw the border–we don't that know yet. As always in science, experiment must have the last word—and investigations are underway! The current view in science is that no real limit, the so-called Heisenberg cut, exists at all. Theoretical formalisms prove this unequivocally. At best, such a limit can be provided by experiment. Meaning: Up to what point are we able to clearly prove quantum effects?"

"You've talked quite a bit about interference now. What's the current world record in experimental quantum interference, so to speak?"

"Oh… One of the records was set in 1999 by Professor Anton Zeilinger and his deserving assistants and PhD students, with whose help and fresh brainpower complicated experimental physics becomes possible. Remember the electrons in the double-slit experiment I just explained? As long as they are not observed, they are in a state of quantum interference. As soon as observation happens, for example when one of the two slits is closed, position information is available, and interference disappears automatically. Zeilinger and his group were able to show something similar with the "smallest footballs" in the world, that is, with the so-called fullerene molecules. These particles consist of 60 or 70 carbon atoms in a pentagonal or hexagonal arrangement. The weight of fullerenes is about 720 atomic mass units. To synthesize them, a furnace was heated to about 600°C. From that, these quantum footballs are ejected with very high speed. Then, they hit a special diffraction grating. There, they have to pass through many small fissures at the same time, so to speak. The statistical distribution of their points of impact was then registered by a certain measuring arrangement. And what do you think the result was? You guessed it. The pattern of stripes, the interference fringes, appeared again. That measurement was enormously complicated because the diffraction angles were so extremely small. Just only that part was someone's PhD thesis."

"Retrospective congratulations to Professor Zeilinger and his team, then. We are curious about how they were able to prove that these quantum footballs exhibit information-related behavior? That they react to observation, so to speak?"

"Oh, I almost forgot! Of course, this was also checked indirectly. No slits were closed for this purpose, though. The furnace was simply heated to more than 1000°C, which was much hotter than before. The scientists found that quantum interference had not completely disappeared, but that it was much weaker. That's because one result of the higher temperature is more heat radiation, i.e. more photon exchange between the fullerenes and their environment. In this way, the molecules reveal more knowledge about themselves, and their observer receives more information. It's a bit like in the story of Hansel and Gretel, where the children lay out a trail of white pebbles, in this way revealing the path they have traveled. In our case, it's not a fairy tale but a science experiment that shows something very interesting. The quantum state of the system obviously depends on its exchange of information with the environment. Even the mutual interactions ('observations') of the atoms in the molecule itself play a critical role. With the increase of this information exchange, their quantum character fades. In technical jargon we would say that the system is getting increasingly decoherent. You can however also see this effect as a manifestation of the uncertainty relation. Because of the higher temperature, the fullerenes have, on average, a higher speed than before. Consequently, the de Broglie wavelength that can be assigned to them decreases. Therefore the position uncertainty increases, which means that the momentum uncertainty has to decrease. The result are smaller diffraction angles and accordingly, weaker interference."

"What was that you said? The de Broglie wavelength?"

"Well, this term can be traced back to a French prince who introduced the concept of the matter wave. There's a related story I would like to tell you. You already know why Albert Einstein received his Nobel Prize, alas the only one he ever got. He was honored for his discovery of photons. At the time, this was revolutionary. Up until then, people had thought that light was exclusively an electromagnetic wave.

Then, Einstein presented an effect which is used today in virtually every solar energy system and in the exposure meter of every digicam, that can only be explained with the concept of light as a quantized particle. With that, Einstein turned the world of physics upside down. A few years later, Prince Victor Louis de Broglie turned the tables again, by assigning a wave aspect to the electron in the atom, which up until then had been considered to be point-like. This wave aspect are the matter waves. De Broglie deduced a formula which states that the wavelengths of material quantum systems, such as electrons, decrease as their mass increases and with that, they become more and more decoherent. In fact, this assumption, which was initially a mere hypothesis, was submitted to Einstein for review. And this is what Einstein said. "He has lifted a corner of the great veil." With his image of a veil, Einstein expressed the then prevalent feeling, even among physicists, that quantum theory was still deeply enigmatic. At that time, neither Einstein nor de Broglie knew that the experimental proof of matter waves had already been produced shortly before by two Americans with their discovery of electron diffraction. Those two, however, had no clue what it was that they had discovered. In any case, de Broglie's bright idea turned out to be a very important step in the development of quantum mechanics. Later, it was taken up by the Austrian Nobel laureate Erwin Schrödinger in his concept of wave mechanics."

"But we humans are also made of matter, atoms and molecules. Is it possible to transfer the concept of matter waves also to humans? Or more specifically, can a human being be assigned a de Broglie wavelength?"

"Actually, that is possible in principle. The resulting number, however, is so small that it lies many powers of ten below the smallest objects known in physics. I am talking about quarks, which together with leptons and gauge bosons form the fundamental building blocks of the world. And it's a good thing that this is so, because otherwise, quantum interference might occur in humans, too. That would be terrible! You see, with an object's increasing mass de Broglie wavelength gets smaller. Compared to a tiny electron, humans have an infinitely larger mass. And so their wavelength is infinitely smaller. Accordingly, diffraction and interference are so infinitesimally small it is impossible to actually measure them.

With that, an important feature comes to the table. We have already come across the term several times. Decoherence. Today, there is considerable scientific evidence that the actual cause of decoherence lies in the exchange of information with the environment. Remember the fullerenes and Hansel and Gretel's path of white pebbles? Something like that happens all the more with humans. We are constantly in interaction with our environment, for example through optical perception, temperature exchange or gravitation. Above all, however, the myriads of atoms a human being is made of are in constant contact with each other. In a way, this also corresponds to something like mutual 'observation'. And all that ultimately leads to the fact that human beings seem to be completely decoherent in comparison to quantum particles, and that's a good thing. So thankfully, we are not subject to the effects of quantum interference."

"What a relief! Thank you very much for your comments! Is there anything else you would like to tell us before we go?"

"Well, remember that there is tremendous potential for technology in the fundamental aspects we discussed. For example, they are the basis for the production of qubits, which can be used for tap-proof quantum communication. Interferometers serve to develop systems where qubits can travel from Alice to Bob with protection that is inherent in the physics. Their safety is primarily a feature of the quantum properties of the system, which in a deeper sense are related to the concept of information. You have seen that this behavior is a universal property. It affects not only light, but all quantum particles, including material ones. Therefore, qubits can also be implemented in the properties of material particles, such as electron spin or nuclear spin. Finally, the phenomenon of decoherence also shows where the main technical challenges of quantum computers and their networking by a quantum internet lie. If we do not want to risk a technical meltdown, it is important to avoid decoherence as far possible, but at the very least to the extent that reliable calculation or transmission is safeguarded."

## 3.2 Phase Cryptography

As mentioned in Sect. 2.6.2, various protocols for quantum key distribution (QKD) exist. We have presented the Ekert protocol, which is based on entangled qubits. It enables a very promising type of quantum cryptography with certain advantages for long-distance communication via quantum satellites. To date, most of the current methods and many of the commercially available systems are based on a connection via fiber-optic cable without entanglement. Typically, they are based on the less complex BB84 protocol. If, however, the qubits are encoded in their polarization, a problem arises in those fiber-optic connections. The qubits' polarization direction is successively twisted by the fibers. An alternative is phase coding, where this effect does not occur. This method is suitable for use in standard fiber-optic cables. It comes, however, with the requirement of much higher interferometric precision. Such systems are also predominantly used in current metropolitan QKD networks.

## What is the Phase of a Wave?

In classical physics, light is described as an electromagnetic wave. If we separate the magnetic oscillation from the electric oscillation (the two are oriented at right angles to each other), all that remains is a simple wave train. Such a wave train can also be developed as follows. Imagine an arrow. Its tip rotates around a fixed center. It continuously draws a circle. Now, mentally project each point of the circle onto an imaginary time axis. If you look at Fig. 3.2, you will see that the result is the model description of a wave. If we now place a second arrow in this model, we get the image of two wave trains superimposed over each other. The two arrows close a certain relative angle $\Delta\varphi$ (pronounced: "delta phi"), which is referred to as the phase angle, or simply the phase of a wave. Two special cases of two superimposed waves are of particular interest. If $\Delta\varphi = 0$, then the waves are "in phase". Wave crest meets wave crest and wave trough meets wave trough. If we then calculate the sum of the respective deflections, their amounts add up to twice that value. We then speak of constructive interference. If, on the other hand, the value is zero, the two wave trains cancel each other out. We call this destructive interference.

## The Asymmetric Mach-Zehnder Interferometer

As successful graduates of Professor Quant's quantum optics workshop, it's no problem for us to understand the functional principle of an asymmetric Mach-Zehnder interferometer (AMZI) (Fig. 3.2). The designation 'asymmetrical' is justified by the different lengths of the U-shaped partial beams. Again, we need to idealize and conceptualize a well-adjusted AMZI. Let's assume that the incident light undergoes a 90° phase shift with each reflection at the beam splitters and mirrors. This means that the relative position of individual wave trains changes by $\Delta\varphi = 90°$ each time. It follows that in this case, only the

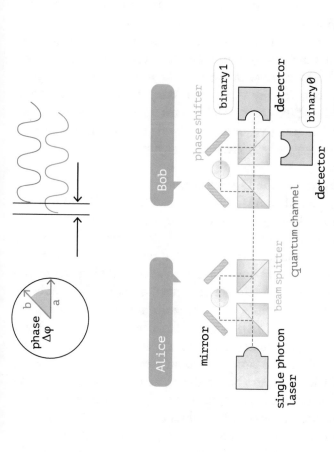

**Fig. 3.2** Large Mach-Zehnder interferometer setup (consisting of two asymmetric Mach-Zehnder interferometers) for phase cryptography. Vector diagram to illustrate the phase relationship.

photon detector "binary 1" will click while the detector "binary 0" will never register a photon. That is, of course, because no photon can ever escape there, as a result of the phase difference between the upper "U-shaped" partial beam and the straight lower beam. That phase difference is 360°—we can say, it's actually 0. The result is constructive interference. Wave crest continues to meet wave crest and wave trough continues to meet wave trough. It's a different story with wave trains which should emerge at detector "binary 0". In that case, the phase difference is $270° - 90° = 180°$, which corresponds to annihilation. This is intuitive, as the entire light intensity has already been assigned to detector "1" anyway.

### Generating the Quantum Key

The described AMZI is now suitable for QKD. In this scenario, Alice and Bob are again two persons/parties/computers who want to send each other messages, in a way that's absolutely tap-proof. What they use is a QKD system that consists of one AMZI each as transmitting and receiving unit. The units are connected to each other, probably with a fiber-optic cable. The entire arrangement (consisting of the two AMZI) is also referred to as the "large" MZI. As photon sources, they use a special single photon laser as well as a quantum randomly controlled phase shifter, which is able to change the phase of the waves according to its setting. A measurement setup for Alice is not shown because she receives her values from the quantum random generator. For maximum safety, however, the generated quantum random sequence must be transferred directly (without buffering) to the automatic phase shifter.

Now, quantum cryptography is performed. The protocol used is the BB84 protocol, which is not unlike the Ekert protocol in its sequence. Steps 1, 3, 5 and 6 can in principle be retained. For steps 2 and 4 (see Sect. 2.6.2), fundamental differences emerge:

- STEP 2: *Generating the quantum key.* Alice and Bob quantum randomly vary the phase. For this, they use a phase shifter (PhS). Whenever the phase difference of PhS is $\Delta\varphi = 0$, the detector must inevitably click "binary 1". If $\Delta\varphi = 180°$, the detector will without fail click "binary 0". If now single photons are emitted from Alice's source, a quantum random sequence of binary numbers is formed. From this sequence, Alice takes a subset to use as the key for the following OTP (see Sect. 2.6.1). In more general terms, it can be said that Alice and Bob quantum randomly generate measurement bases which are then publicly communicated in order to delete measurements with different bases. In those cases where the basis is the same, one speaks of the relevant bits, in contrast to the irrelevant bits, where the phase relationships outlined above do not exist. In the latter cases, it is not possible to predict with certainty which detector will respond. This is because the emission from the beam splitter underlies quantum randomness. To make this distinction, Alice and Bob use a classical channel to communicate exclusively the number of the respective photons and the corresponding bases (never the relevant bits).

- STEP 4: *Espionage test.* The detection of an eavesdropping attack is done by statistically evaluating the generated list, which depends on the respective type of implementation. To achieve this, Alice and Bob take a sufficient number of test bits and check whether they match. They will, of course, then delete those bits from the actual key. Once again, we need to consider the purpose of QKD. The goal is not to transfer information directly via the single qubits, something that's not physically possible anyway, but only to generate and assign an original, absolutely random quantum key from Alice to Bob, which will then be used for subsequent data transfer via OTP over the normal internet. The inherent security is guaranteed by the fact that any hacker attack by Eve would be detected automatically during the key transfer. If this happens, Alice and Bob can discard the key before any data has been transferred. In a QKD network, an alert would be given on the key management level and a new route would be offered, or already generated secure quantum keys would be provided at short notice.

**Detecting an Eavesdropping Attack**

The inherent security is based on the validity of the no-cloning theorem. For an eavesdropper (Eve) to gain knowledge of the quantum key, the entire quantum state would have to coexist with Eve independently of Alice. According to the no-cloning theorem, however, this is impossible. The reason is that the quantum state cannot be copied perfectly. In practice, this leads to the fact that every single one of Eve's measurement attempts influences the entire quantum state in such a way that this becomes noticeable by evaluating the measurement statistics. An example: Alice and Bob vary their phase shifters on two different measuring bases. One base is the axis "0–180°" in the vector diagram, a second base forms the axis "90–270°". By rotating the phase shifters completely randomly, Bob would on average measure 50% in the correct base (relevant bits), but 50% in the wrong base (irrelevant bits). This is because $\Delta\varphi$ is unequal to 0, or 180°. Suppose Eve attempts to eavesdrop on the transmission. Eve is even furnished with insider knowledge and knows both bases. Furthermore, she uses the same AMZI for phase measurement and she is able to forward the received bits to Bob with extremely high speed. Even then Eve would have no chance. Due to the quantum random modulation, which is impossible to know in advance, Eve would measure in 50% of all cases in the right base, but 50% of all cases in the wrong base. The latter would inevitably influence Bob's measurement. He would then measure only 50% of the bits sent by Eve as matching Alice's bits. His overall error rate would be 25%. Alice and Bob would immediately notice this upon comparing their values. Theoretically, Eve could attempt to measure only every second or third photon. The error rate would then be reduced to 12.5%, or 6.25% respectively, and so on. At some point, the error would remain undiscovered. But even that is of no use to Eve, because she also loses information about the key successively to the same extent.

### Decoy-State QKD

In contrast to the principle outlined above, a perfect single photon source does not exist in real life. Therefore lasers with very low intensity or very weak coherence are used in practice. However, multi-photon states also occur, and this limits the secure transmission rate significantly. For example, an eavesdropper could use a beam splitter to remain unnoticed while measuring individual photons as they are being generated. We do not want the transmission rate to degenerate into a lame duck while we want to safeguard security. This problem is frequently solved using decoy states. Instead of a coherent laser beam, Alice uses laser pulses with different intensities (one signal state and several decoy states). As a result, the photon number statistics varies throughout the channel. Alice then publicly reports to Bob the intensity level used on each qubit. Any eavesdropping attack is detected efficiently by measuring the error rate (QBER) of each level. Current test QKD networks predominantly use such system. The safety gain due to the decoy-state method is scientifically proven.

# 3.3 Schrödinger's Cat

The thing about cats. One moment they curl up in our lap affectionately, purring amiably. The next moment they hiss wildly, scratching and biting, suddenly transformed into furry little devils. It seems that two opposing principles are at war within our cats, much like two opposing quantum states. Maybe that's why a cat had to serve for one of the most famous scientific gedankenexperiments in the world.

At a time when the currency of the euro was not yet created and Austrians paid for their purchases in shillings, the 1000 shillings banknote of that era was embossed in blue hues with the engraved image of a man with the receding hairline of a scholar. That man was the quantum physicist and science theorist Erwin Schrödinger. Naturally, only high merit will lead a country to stamp your face onto a banknote. And indeed, Erwin Schrödinger was the father of the wave function and the creator of wave mechanics. His contributions to quantum physics are beyond dispute.

Let's go back to Newton's principle of action. With this equation of motions, every track curve of any classical object imaginable can be calculated. However, Newton's formula fails completely when it comes to nonclassical questions about atoms. One example is the "orbital curve" of an electron around the atomic nucleus. This is on the one hand due to the fact that because of the uncertainty relation, no defined trajectory curve exists at all. Therefore, this notion is essentially meaningless. On the other hand, however, classical electrodynamics is relevant for such phenomena. A rotating and therefore accelerated charge such as the electron's generates an electromagnetic wave. In order to exist, this wave has to permanently draw energy from its environment. The logical consequence would be that the electron falls into the atomic nucleus in a spiral trajectory. How atoms can remain stable was inexplicable for scientists at the time. This was only one of a large number of problems that kept theoretical physicists of the time busy.

Erwin Schrödinger brought the atom problem to a surprising solution by coming up with two ingenious ideas. Firstly, he introduced the abstract concept of the wave function. And secondly, he wedged this $\psi$-function into the structure of a so-called eigenvalue equation. The result was the world-famous Schrödinger equation, which today is one of the most cited formulas in physics. To understand it even rudimentarily, think of qubits in a quantum computer. There we represented the quantum bit as a linear combination of fundamental states, which upon measurement always decay into the eigenvalues 0 or 1. Such mathematical structures, originally derived from linear algebra, were taken up by Schrödinger and successfully applied to the atom with the aid of functional analysis. Similar to the qubit, eigenstates and eigenvalues can also be calculated for atoms and molecules, but in a much more complex form. The eigenstates (eigenfunctions), whose amplitude squares were interpreted by Max Born as probabilities of physical measured values, today form the basis of modern chemistry and are commonly known as orbitals. The associated eigenvalues correspond, for example, to the quantized energy levels in atoms. In quantum mechanics, observables are generally assigned to Hermitian operators, whereby the eigenvalues of the associated eigenfunctions correspond to real measurement values. Mind you, wave mechanics was not the first mathematical formulation of quantum mechanics, the equivalent, Heisenberg's matrix mechanics, was developed a little earlier. However, the Schrödinger equation is generally regarded as less cumbersome, as it considers operators and wave functions in a single equation of motion for the states, whereas in matrix mechanics, equations of motion stand for the operators themselves. Later, the Schrödinger equation was modified and developed further. The English theorist Paul Dirac combined it with the special theory of relativity, which led to a sensational discovery: the existence of antiparticles such as the positively charged electron (positron). This heralded a development which was later denigrated as a "particle zoo", but today forms the basis of all modern physics. In 1933 Schrödinger and Dirac jointly received the Nobel Prize in Physics.

### The Cat Paradox

In public, Erwin Schrödinger is better known for a certain world-famous metaphor. Every child has heard of Schrödinger's cat. But what is all that about? Professor Quant has made us familiar with the term wave function. What does the wave function have to say about superposition states, i.e. the phenomenon of superposition? As explained, in quantum physics it describes the most diverse states: quantum parallelisms in quantum computers, superpositions in interferometers, interferences in atoms and fullerenes and so on and so forth. Does this function exist everywhere, even in larger, so-called macroscopic systems? Perhaps even in humans? Professor Quant has also already given his point of view as a physicist. Actually, this question was raised many decades earlier by Erwin Schrödinger in a journal article published in 1935. Probably out of his sense of tact, Schrödinger did not use a specimen of the human species in his essay. Instead, he cast a cat in the leading role. In any case, the living organism was central to the work interest of this exceptional physicist. Among other things, he wrote a widely acclaimed book where he addressed questions about the origins of life. There, he anticipated human DNA even before it was actually discovered by Watson and Crick shortly afterwards. Erwin Schrödinger was not only a brilliant scientist, but also a writer with exceptional eloquence and style. The aforementioned "cat article" is therefore reproduced in extracts here (translated into English by John D. Trimmer):

> (…) One can even set up quite ridiculous cases. A cat is penned up in a steel chamber, along with the following device (which must be secured against direct interference by the cat): in a Geiger counter there is a tiny bit of radioactive substance, so small, that perhaps in the course of the hour one of the atoms decays, but also, with equal probability, perhaps none; if it happens, the counter tube discharges and through a relay releases a hammer which shatters a small flask of hydrocyanic acid. If one has left this entire system to itself for an hour, one would say that the cat still lives if meanwhile no atom has decayed. The psi-function of the entire system would express this by having in it the living and dead cat (pardon the expression) mixed or smeared out in equal parts. It is typical of these cases that an indeterminacy originally restricted to the atomic domain becomes transformed into macroscopic indeterminacy, which can then be resolved by direct observation. That prevents us from so naively accepting as valid a "blurred model" for representing reality (…)

Let us go back to Professor Quant's Mach-Zehnder experiment (Fig. 3.1) and transfer that situation to the cat paradox. A photon that emerges from the source has a quantum random probability of 50% for being either transmitted or reflected at the first beam splitter. The situation is similar for the cat paradox scenario. Here, too, there is a 50% chance for the cat to be either dead or alive. This depends on whether the atom decays or not, which statistically happens with a 50% probability for each case. In the Mach-Zehnder interferometer, a superposition of both possibilities exists. We recognize this when we see the interference fringes. This could be expressed in wave function in the following way:

$$\Psi = \psi_{transmitted} + \psi_{reflected}.$$

In analogy to this, Schrödinger now with great seriousness asks the question whether the state of the cat in the steel chamber should not also be determined by a superposition state, i.e.

$$\Psi = \psi_{cat\ is\ dead} + \psi_{cat\ is\ alive}.$$

He does this with a sense of humor, a little reminiscent of Einstein's "spook", and just like Einstein, he wants to show that the entire situation contradicts our everyday experience, that it is paradoxical. Nobody has ever seen the interference image of a cat, and certainly not that of a human being. But because quantum formalism theoretically permits this grotesque situation, Schrödinger indirectly raises a number of further questions in his essay:

1. Is the formalism of quantum mechanics also valid for macroscopic objects?
2. Is there a defined limit, beyond which it would no longer be valid?
3. What role does the measurement process play in the Copenhagen interpretation of quantum mechanics?

### The Explanation, According to Most Physicists

In fact, it is still not possible to make a science-based decision, with any certainty, about the paradox of Schrödinger's cat. The thought experiment offers scope for numerous interpretations, some of which are quite bizarre, but also for concrete experiments that recreate the scenario in a modified form. However, a plausible and quite elegant explanation does exist, and the majority of physicists agree with it. The superposition of dead and living cats is only possible if the animal, but also the hellish machine trapping it, are treated as quantum objects. If that is the case, however, the experiment becomes an impossibility. The reason is that an "inner" measuring process occurs, which renders all further measurements entirely decoherent right at the outset of the experiment. The cat and the measuring setup immediately become classical objects and are automatically no longer subject to quantum theory. This internal measuring process can be interpreted as the exchange of information with the environment. The predominant contributors to this are the myriads of atoms and molecules the cat is made of, but also the experimental equipment, and the box itself, and the air inside the box. In principle, there exists yet another take on the situation. The Copenhagen interpretation of quantum mechanics assumes as an essential requirement that reality does not exist prior to any measurement. Any reality is only determined by the nature of the measurement process. If the atomic decay and the cat really are entangled quantum objects, then life and death would actually be indeterminate. Only by opening the chamber (which corresponds to the measuring process) would one of the two states become reality. As a result of decoherence, however, the measuring process is reduced to atomic decay only, which then has to be completely separated from the cat. The measurement process itself is arbitrarily assumed to be a classical mechanism.

## Why Do We Experience the World as Classical?

The paradox of Schrödinger's cat, then, is that the quantum formalism allows superposition and entanglement everywhere, but we aren't able to notice any of that in our everyday world. None of the cats we see, or any other objects we interact with, seem to exist in strange superposition states or blurredly appear to be in several places at the same time. Quite the contrary. Everything has its fixed location, clear contours, definite speeds and distinct directions of movement. This classical perception seems to be at odds with the bizarre quantum world. We would be tempted to think that quanta are just a special case of classical physics, which occurs only in the tiniest detail. Actually, the exact opposite is the case. Our everyday world turns out to be a special case of quantum physics. This always occurs when decoherence effects play a major role. Just how efficient decoherence is can be seen from the fact that every bacterium is already a classic object, even though it is so tiny that it remains invisible to our human eyes.

So, how exactly does decoherence emerge? In fact, the mathematical formalism predicts that macroscopic superpositions will emerge from microscopic superpositions. What happens, however, is that the state of the system gets entangled with the state of the environment. The result is a nonlocal quantum state. By exchanging information with the environmental degrees of freedom, the macroscopic superposition becomes delocalized. This process can be described mathematically with a multidimensional configuration space. From our low-dimensional perspective, however, the world appears to be localized, isolated and therefore classical.

What is important to note is that quantum interference disappears locally, but not generally. Our world appears classical only from the perspective of a local observer. The slightest interaction (for example with light, the scattering of air molecules, etc.) already starts the decoherence process, as can also be shown mathematically. In consequence, the world we perceive must inevitably appear classical to us. The fact that it really is not (as many theorists think) and instead has a nonlocal character (which possibly extends through the entire universe) is quite suitable as a new paradigm of natural philosophy.

**Cat States in Quantum IT**

Decoherence also addresses a fundamental technological problem in the development of a quantum internet. Macroscopic objects inevitably couple to the degrees of freedom of the environment. How pronounced this effect is depends on the nature of the environment and the interaction with it. For microscopic objects such as atoms, decoherence is usually so weak that their behavior has a strong quantum character. For larger molecules, on the other hand, this coupling is much more noticeable. (That's why it was so difficult to detect quantum interference in Zeilinger's fullerenes.) The most complex molecule known in the universe, human DNA, is already so decoherent that its behavior is mostly classical. Lucky for us, because otherwise, heredity would not be possible at all! For these reasons, quantum devices have to create conditions so exotic that the exchange of information with the environment is consistently prevented. This applies especially to quantum computers. Previous implementation attempts, which seem somewhat surreal, have demonstrated that it's all about systems where decoherence can be prevented as well as possible. If you have ever wondered what ultrahigh vacuum or ultracold temperatures near absolute zero are good for, here is your answer. Precisely those environments are able maintain coherence as well and as long as possible.

Another category of quantum processes is represented by so-called macroscopic quantum states, which, for example, suddenly start when a certain temperature is reached. Such macroscopic quantum states include superconductivity or Bose-Einstein condensation. What we are looking for are systems that are as well isolated as possible and at the same time can be manipulated, with promising coherence properties. This is the most important challenge for the quantum internet, but above all for the quantum computer. It's still not possible at the moment to implement such systems in a way that a sufficient number of qubits in the group maintain their coherence, so we are able to perform interesting operations with them. This is among the most important development challenges today. We are still not able to provide definite assessments with regard to performance and scaling. As mentioned, however, the paradigm is precisely that the world is "coherent" by nature. One of the most exciting questions therefore is how much of that can actually be used for technological applications. Nature per se seems to be calibrated exactly in this way.

## 3.4   Workshop: Beaming People—Is that Possible?

In our times, hardly any physics concept has fascinated human imagination as strongly as the idea of beaming, or more precisely, quantum teleportation. This strong interest is sparked above all by the question: Will we ever be able teleport people from A to B, like they do in the sci-fi series "Star Trek"? The developers of the quantum internet hardly devote any attention to this. Nevertheless, we might ask: What if…? We return to our fictional Institute for Quantum Optics, where Professor Quant will introduce us to theory and practice of these mysteries.

"Welcome, welcome! I am glad you'll join me for another workshop. This time it's about one of the most exciting things in physics: the beaming of quantum information. You are aware that this has nothing to do with what you may know from science fiction. What we will do is, we will transfer specific quantum information from one object to another. That other quantum object exists already, and it is spatially separated. So we don't actually transmit matter. All we transfer are certain physical properties. What's especially amazing about this is that the state that's to be transmitted does not even have to be known to the experimenter. We are able to do fly blind when we do teleportation, so to speak. So you see, teleportation in physics is fundamentally different from beaming in science fiction. The latter was only invented to lower the production costs of such films... Let's take a look at a rather straightforward teleportation experiment. We will discuss possible questions or FAQs afterwards."

Professor Quant leads us into a mysterious dark room which is lit dimly only by strange laser light. On a large table we see an assortment of mirrors and other devices. Suddenly two figures appear out of the semi-darkness. They smile, and they appear to be friendly. "May I introduce," says Professor Quant, "my two assistants: Alice and Bob. They will now perform photonic teleportation."

The professor takes us to a whiteboard (Fig. 3.3), where graphs seem to illustrate the theoretical background of the experiment. One of the drawings somehow reminds us of the Mach-Zehnder interferometer from the last workshop. "That's right!" the professor says approvingly. "This is a highly modified form of the interferometer. In addition to a nonlinear crystal system, we are using a polarizing beam splitter, which allows us to also measure the polarization when we combine it with two detectors (those are not shown). That's something a normal beam splitter cannot do. My assistant Alice now starts the teleportation. She creates two EPR pairs using parametric down conversion. She then prepares one of the four particles with the state to be teleported and performs a Bell measurement with one of the particles of the second EPR pair. Bob, for his part, measures the photon of the EPR pair assigned to him. In this way, he obtains the quantum state of Alice's photon. That's how quantum information is beamed from

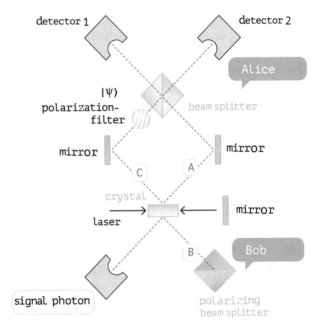

**Fig. 3.3** Photonic teleportation

Alice to Bob. The interferometer in our design here needs to be extremely precise, and that's not possible outside the clean and controlled laboratory environment. So, we use a combination of a fiber coupler and a beam splitter based on the quantum mechanical tunnel effect and…"

"I beg your pardon, Professor, but could you explain this more clearly, please?"

"Uh… of course. I will explain each step to you again, very slowly.

Let's start with parametric down conversion. You already know this term from quantum cryptography. A blue laser pulse passes the crystal and generates a pair of red photons, which are entangled in their polarization. Alice (A) receives one of them, the other one goes to Bob

(B). The pulse is reflected back by the mirror and passes through the crystal again, where it creates a second entangled photon pair. One of the two entangled photons serves as the signal photon (it indicates that a photon is ready for teleportation), the other one goes to Alice (C). That one passes through an adjustable polarization filter, where the state that is to be teleported $|\Psi\rangle$ is created. For historical reasons, the entangled photons are named after the physicists Einstein, Podolsky and Rosen (EPR)."

"How can a single photon from a blue laser produce an entangled pair of red photons?"

"First of all, this is possible because a blue photon has twice as much energy as a red one. The reason is that the frequency of the red light is only half that of the blue laser light. For reasons which are terribly complicated, I'll spare you the details, the photons from the crystal fly off into separate directions making the shape of two cone shells. In those locations where the two cones overlap, it's no longer possible to distinguish the particles from each other. In a way, information is lost there, and that is an important prerequisite for entangled states."

"In what capacity are those particles entangled?"

"They are orthogonally polarized to each other. This means that if, for example, one of the photon pairs is measured with respect to its plane of oscillation, the plane of the entangled partner particle is automatically set at right angles to it."

"How exactly does the Bell measurement work?"

"Alice brings the two photons A and C together in the beam splitter. Then, she measures them with detector 1 and detector 2. If you think of the measurement, remember our Mach-Zehnder interferometer from the first workshop. When we inserted polarizing filters into the beam paths, we observed that interference disappears immediately in those cases where the polarizing filters are orthogonal to each other. Inevitably, however, this also means that photons A and C have to be polarized orthogonally. Now we have to remember that in this case, both detectors would click in a well-adjusted interferometer (which we assume here). It's a different story with interference, where only one of the two detectors would respond. That's an unambiguous measurement criterion."

"And what does that have to do with the actual teleportation?"

"That follows logically. So, Alice measures one photon at detector 1 and one at detector 2. She now knows for sure that A is orthogonal to C. However, because of entanglement, A is at right angles to B. It follows that states B and C are identical. Now Bob measures this polarization state with his polarizing beam splitter and two detectors behind it (not shown). The entanglement disappears, and the teleportation is completed. And that's how the state created by Alice $|\Psi\rangle$ was transferred to Bob's photon B."

"Fair enough. Bob now has a photon that is in the same state as Alice's. What's that got to do with teleportation? It's like taking a piece of paper, drawing Alice's state on it and faxing the paper to Bob. Then why don't they call that quantum faxing?"

"Indeed, that idea would make perfect sense", Professor Quant explains, "however, to fax something means to reproduce information. But that is strictly prohibited in quantum physics. That's not what happens in our experiment. Note that the quantum state of Alice's photon disappears the instant it appears in Bob's photon. So the quantum information is not copied, but rather transferred to another, locally separated quantum object (in our case that's a photon). This, in essence, is the no-cloning theorem, another fundamental principle of quantum mechanics."

"But what if Alice and Bob measure both particles at the same time? Then you would end up with the same state twice, and the info would have been doubled. Einstein would have his EPR objections, but the no-cloning principle would be refuted…"

"Well, no—because fortunately, that same Einstein discovered the theory of relativity, which states that absolute simultaneity cannot exist. Meaning that we could always find a reference system where the events are not simultaneous. And there, everything would be fine. Believe me, from what we know today, the no-cloning principle is very well compatible with teleportation. To this day, nobody has ever been able to produce scientific proof that would refute the no-cloning theorem."

"But how can the particle remain an original during teleportation, how can it preserve its identity if the object itself is not transmitted at all?"

"This brings us to the question of what identity or individuality of something is at all.

Let's take humans as an example. Every human being consists of perhaps $10^{28}$ atoms. That's an immeasurable number with 28 zeros. You, me, Alice, Bob, we are all made of the same kinds of atoms, mainly carbon and hydrogen. Our material composition is the same. So what, then, makes our individuality? The answer is quite simple: the *how*. For example, the way these atoms are arranged. This means that what we are in the real world and what makes us an original is the information about the properties of our atoms, but not primarily matter itself. It's exactly this kind of information that is being transmitted in teleportation. As I said, there is no difference between information and individuality. Therefore, we can say that the original was actually teleported."

"Isn't there a certain contradiction to Einstein's theory of relativity? After all, quantum teleportation takes place at a velocity beyond the speed of light. The theory of relativity expressly forbids such a thing, doesn't it?"

"Oh, you have to consider what the theory of relativity really says. It does not claim that there can be nothing that moves faster than the speed of light, but that the information we can use must not be transmitted faster than the speed of light. In fact, this principle is also applied to quantum teleportation, since Bob can never know exactly whether the teleportation was successful. He has to ask Alice, for example, and she could at best give him the answer at the speed of light. In technical jargon, we could express it like this: "For this information, Bob has to leave the quantum channel and switch to a classical channel.""

"And why does Bob have to ask Alice if the teleportation was successful? Besides, why ask? If she were to stand facing him, like she is doing here, she could just give him a little wave or something."

"Sure, he could visually perceive her—or the detectors. For this, at least one photon that is reflected by Alice or the detectors would have to arrive in Bob's eye. And if there's anything that cannot go faster than the light, then it's light itself!

Let me come back to your first question. Do you remember Dr. Bertlmann's socks? If they were quantum socks, their color would be completely undefined, until John Bell (or whoever) looks at them. Only then will the socks take on one of their possible colors. We can't know the color before we see the sock, in principle. The same is true for our experiment. We have four different possibilities, so-called Bell states. And all of them are objectively random. Only one of these four states enables teleportation. Since all states occur with the same probability, teleportation can on average only occur in 25% of all cases. So, Bob can never know for sure whether the measurement actually was a teleportation or not. Only when he receives that information from Alice does he know with certainty what the status is. For this, he needs a classical channel."

"Okay then. It really was teleportation. If teleportation isn't the transportation of concrete quantum objects, but rather the transfer of pure information—the information that characterizes the original, if we assume that a human being also consists of a certain amount of information, and that it must be possible to transmit this information somehow, shouldn't it be possible at some point to teleport humans?"

"I'm afraid I have to disappoint you! According to today's assessment, teleporting people is absolutely impossible!"

"But why? You just said that humans are pure information, so to speak, and information can in principle be transferred."

"That's exactly what's at the heart of one of the unsolvable problems. It's not possible to read the entire information of any person in our world. The reason is Heisenberg's uncertainty principle. Think about the $10^{28}$ atoms a human being is made of. You would have to determine their exact arrangement. For that, you would have to know the atoms' exact position—and their momentum. According to Werner Heisenberg, this is impossible to know position and momentum of a particle at the same time."

"But why then can you do this with the photons in your experiment? Aren't they also subject to Heisenberg's uncertainty principle?"

"They sure are. But even in our experiment here, we don't transmit the entire information of a photon. Keep in mind that the particles are only entangled in their polarization—and only this information is transmitted. Theoretically, the complete information can only be transferred if it is absolutely unknown to us. And of course that's no help, because the moment we want to get at this information, we have to make a measurement, and that automatically disrupts the system."

"If this is the case, we do not understand why teleportation works at all. On the one hand, it is based on entanglement, which would have to be destroyed as soon as Bob measures the particle state. No entanglement—no teleportation, one would think."

"Very perceptive. Let's take a closer look. In contrast to the Mach-Zehnder interferometer in the first workshop, where there was only one photon in the interferometer, here, we find *two* photons. This changes the situation considerably. Detector 1 and detector 2 only determine whether the two photons are orthogonally polarized to each other, but not which of the two photons has exactly which polarization. In this respect, the detectors do not reveal the full information. We are left with a certain residual entanglement, which is sufficient for teleportation to occur. This entanglement rest is only destroyed the instant Bob carries out his measurement with the detectors (not shown) behind his polarizing beam splitter. Only then does it become clear which particle carries which polarization."

"Wouldn't it then be possible to at least transmit parts of a person's information, as is done with those photons? To carry out something like a partial teleportation?"

"You really are persistent! Okay then, even if we disregard the problems with Heisenberg's uncertainty principle we would fail in the practical implementation. How are we supposed to entangle $10^{28}$ atoms when currently, we are able to manage just a few? And even if we could, how exactly would this entanglement work? To entangle something means, in principle, to make it indistinguishable. What we do is, we create a situation where we lose information. We can also call that quantum coherence. How are we to produce this situation with a macroscopic object like a human being? Think of Schrödinger's cat! If instead of a cat, we were to place a person in an extremely well-protected steel chamber, they would immediately become decoherent as well. Only just the internal thermal interaction, that is, the inevitable exchange of information between the myriads of atoms that the person is made of and those in the environment, would cause this decoherence. And that's without even considering other things, such as gravitation, something we cannot simply switch off. Also, if we wanted to perform teleportation in the way we are doing with photons here in our lab, there's another impossible challenge. The person to be teleported would have to be entangled with an EPR twin pair. It is completely unimaginable how this might happen in a human being or what that actually means. No, no. The idea of beaming humans remains firmly in the realms of science fiction."

"What, then, is the significance of quantum teleportation?"

"Above all, the significance of quantum teleportation lies in the future of technology. You see, as a method for connecting quantum computers in a quantum internet one day, quantum teleportation is very promising. It has long been proven that individual qubits can be teleported. Also, that teleportation over long distances is possible. A Chinese team was able to achieve the remarkable distance of 1200 km. The arrangement in our lab can be done in a much more complicated way with a quantum satellite! In that scenario, Alice and Bob are very far away from each other. As you know, it is possible to realize quantum repeaters by exchanging entanglement, or entanglement swapping (which corresponds to a teleportation of entanglement). With this, the most important requirement of a quantum internet is fulfilled. You see, in our comparatively simple experiment here, all that's transmitted are the polarization states of individual light quanta. The same principle can theoretically be applied to much more complicated entangled states. Perhaps one day, researchers will succeed in creating and preparing such extremely complex states. We could then teleport those as the output from one quantum computer to another, where they are used as input. Who knows what the future holds and which technological applications will become reality? 3D printers, for example, are already available today. I am able to download complex data from the internet, and the printer then produces a three-dimensional structure from that. Imagine a quantum device that downloads complex quantum information from a quantum cloud via teleportation to an existing medium. In principle, quantum computers are able to simulate enormous atomic and molecular structures. It might therefore become possible to create an exotic material that adapts to its user's wishes. That would be certainly be very convenient, because in addition to its external shape, we would also be able to synthesize the corresponding material, more precisely its quantum properties, as desired. We would be able to create designer material for a wide range of applications. All this, of course, is still in the realm of speculation and fiction. From a physics perspective, that's in principle conceivable. In any case, quantum teleportation is an extremely exciting phenomenon, and we are far from knowing all the applications we could use it for."

## 3.5   A Journey Into the Future

Not only quantum physics itself plays a role for the quantum internet, but also Einstein's theory of relativity. Indirectly, the theory of relativity also guarantees its inherent security, above all because it supports the existence of the no-cloning theorem. The following two sections are intended to make this connection understandable. We start with a brief crash course on special relativity theory. Let's assume that super engineer Gyro Gearloose invents the ultimate rocket that breaks all dimensions and is able to reach extremely high velocities. Now Huey and Dewey place their brother Louie into that spaceship. The vehicle with Louie in it leaves Earth. After a few months, it accelerates and almost reaches the speed of light (about one billion km/h in a vacuum) before returning to Earth. When Louie exits the rocket, however, he is in for a surprise. His two brothers Huey and Dewey, who like himself were children when he set out for the stars, are suddenly as old as he remembers their great-uncle Scrooge McDuck to be. How is this possible? Is the rocket a time machine? Louie has obviously traveled into his brothers' future—while he himself remains (almost) as young as when he left.

Cross your heart! You have probably heard of this paradox, be it with twins, triplets or watches. It's not easy to believe, though, that something like this could really happen. In fact, the watch paradox really is pure science fiction, but only as far as market-compatible passenger transport is concerned. It's a different story with elementary particles. And it has already been explained that in physics, the laws of nature are only accepted once they have been sufficiently verified by experiment. So, let's take a look at the relevant experiments.

In fact, the triplet paradox is only the caricature of a phenomenon that belongs to the "core business" of relativity theory, an effect that's referred to as time dilatation. It is, of course, not practically possible to perform such an experiment in the way described above. But if the astronaut Louie gets much smaller, if we were to shrink him, for example, to the size of an elementary particle, it becomes easier to accelerate him to appropriate, i.e. "relativistic" velocities. It's actually quite easy to observe such "time travel" in nature. One example are muons, particles which are formed about 10 km above the earth's surface when the high-energy particles of cosmic rays collide with air molecules. These muons race down to the earth's surface almost at the speed of light, and there, we can measure them. What makes the whole thing more interesting: The average life span of muons is extremely short. Their decay time is about 1.5 millionths of a second. Even at the speed of light, they aren't able to travel more than half a kilometer in that time span. The solution of that riddle is really quite simple. Each muon flies—just like Louie in our triplet paradox scenario—into the future. Or, to put it another way: Relative to Earth time, the so-called eigentime of the muon slows down dramatically. So the time that passes for a muon on its journey is much shorter than the time that passes meanwhile for us earthlings.

For this effect, this time dilatation, another interpretation is possible. If we depart from the perspective of terrestrials, for whom the muons' time passes slower than their own, we can adopt the muons' point of view. Seen from the muons' perspective, Earth whizzes towards them almost at the speed of light. For that case, the equations of relativity show that our planet would no longer be spherical, but strongly flattened, because all distances in the direction of motion (of the planet!) would be shortened. For the muon, the distance to Earth's surface then is less than half a kilometer. That's a distance it can travel easily within its lifetime, and still be detected by physicists on Earth. This shortening of distances of moving objects is called length contraction.

The generation and the arrival of the muon on Earth's surface are two physical events which are measured as time dilation from the terrestrial point of view, but from the muon's perspective are seen as length contraction. And this is precisely at the core of relativity theory: The appearance of nature depends on the observer's point of view, or more precisely, on the state of motion of his reference system, provided it is a so-called inertial system. The special theory of relativity considers inertial systems only. The question of who "really" is right is quite pointless. For in the universe, neither absolute time nor absolute space exists. Moreover, space and time are closely connected and can be proven to be as flexible as a rubber band.

So, what's an inertial system? Imagine a bowl of soup that's sitting on your kitchen table, waiting to be eaten. The surface of the soup is completely smooth—until we dip our spoon into it. The same would be true if the bowl were sitting on the table of a comfortable airliner that's moving along completely calmly and uniformly. What's the difference? If one did not know that the soup was on an airplane, its behavior would not be distinguishable from the bowl of soup on the kitchen table. In the first case, the speed of the soup relative to Earth is zero, in the second case it may be 850 km/h. This is exactly where we come across one of the two fundamental assumptions of special relativity. In uniformly moving reference systems without windows, their occupants cannot distinguish between "rest" and "movement". This is why all inertial systems have to be regarded as equal. This assumption leads to highly interesting consequences, namely the relativity of simultaneity. Imagine a flight attendant at the back of the plane and another one in the front area of the plane. Exactly in the center between the two, someone produces a flash of light with their digicam. Photons are emitted, which reach the eyes of both flight attendants at exactly the same moment. They both perceive this event at the same time. From the point of view of an observer on Earth, however, the conclusion is quite different. From their point of view, the photons reach the flight attendant in the back earlier than the one in the front area. This is because the tail of the airplane flies towards the light beam, while the nose tries to fly away from the photons, so to speak. So if seen from Earth, there is *no* simultaneity. Who is right now? Both are right! In view of the equal rights of inertial systems, no absolute simultaneity of events exists. If in an inertial system two events occur simultaneously at two different locations, these events occur at different times take place in an inertial system that moves relative to the first system. That's what the principle of relativity is all about.

Of course, time dilatation is not only a phenomenon of inertial systems. It also occurs in accelerated systems and even in everyday life. We just don't notice them. When Lewis Hamilton gets out of his race car after a grand prix, he is actually a little younger than the spectators in the stands who, relative to him, are at rest. The only reason he does not notice that is because that time difference is very small. If, however, he had an extremely accurate atomic timekeeping device on his wrist (if atomic clocks were to exist in wristwatch sizes), he would be able to determine the tiny time difference exactly. In fact, such experiments were carried out with high-precision atomic clocks in airplanes and satellites. In this way, the effect of time dilation was scientifically proven very directly. So, in everyday life, we encounter minimal relativistic effects all the time. It's no problem for us to neglect them, because everyday speeds are so much lower than the speed of light. However, this does not fully apply to technical applications. Did you know that the navigation system in your car only works correctly because it takes time dilation into account? Among other things, this technology is based on the comparison of runtime signals of electromagnetic radiation. Since satellites—remember the quantum satellite, where the same principle applies—orbit Earth at considerable speeds, time passes a little differently in your car than on the satellite. High accuracies are required for our GPS system, and so this deviation has to be taken into account. If it weren't, our navigation device would be hundreds of meters off!

Let's return to the triplet paradox. One objection that's often raised is that we cannot even know which one of the two observers is moving and which one is at rest (and therefore we might innocently question the time dilation effect). Louie in his spaceship of course feels it's not him who is the moving observer. He feels himself to be at rest. From his point of view, Earth first races away from him and later again towards him. This, in fact, can be invalidated by looking at the practice. Louie has to depart Earth. His spaceship needs to accelerate in order to get even close to the speed of light. Then, however, his spaceship is no longer an inertial system, which must not be accelerated. Therefore, this system would be considered as excellent compared to the other and would no longer have equal rights. In addition, in practice it is necessary to consider the reversal of the space vehicle's path, which in turn requires acceleration (this also holds true in the negative sense, i.e. braking). Although an exact analysis is indeed quite tricky, one definitely arrives at this conclusion: Time has passed much more slowly for Louie in his spaceship than for his brothers on Earth.

Nevertheless, the principle of relativity on its own does not explain exactly *for what reason* time dilatation occurs. It creates a prerequisite but does not explain the phenomenon quantitatively. This basic assumption therefore has to be supplemented by a further axiom which is quite essential. Einstein initially formulated this as the principle of the constancy of the speed of light, but later extended it in favor of a much more fundamental assumption: the signaling effect. It says that useful information (i.e. information that is concretely accessible) can never be transmitted faster than the speed of light in a vacuum. Let's take a look this effect has on our astronaut Louie.

## The Ticking of the Biological Clock

Imagine Louie roaring along in his super rocket. See how he tilts his head a little to scrutinize the tip of his toe (T).

For him to see his toe—to receive optical information—at least one photon has to pass from T to Louie's eye (E). And for that, it obviously has to cover the distance TE. From the point of view of the stationary inertial system on Earth, however, the light quantum has to cross the much longer distance (TE)′ because the spacecraft moves at very high speed during this time. The decisive element now is that the photon, which transmits useful information, can never exceed the speed of light, in agreement with Einstein. It therefore takes the photon a longer time to cover the distance (TE)′ than the distance TE. This means that the information about Louie's toe is received by his owner in the racing spaceship at a later point in time than if he were standing on Earth's surface and looking down on his feet there. For this reason, the speed at which time passes for a moving object relative to a resting observer must inevitably decrease, in order for the velocity of light to retain the same from both perspectives (photons per se always have speed of light). With this, we can explain time dilatation also quantitatively. We are now able to calculate it directly from the relative speed and the speed of light respectively. This effect starts with any relative velocity that does not equal zero. However, significant deviations occur only as we approach the speed of light. That's why it remains hidden in our everyday world. Now, what does this mean for Louie's biological age? After all, he has traveled into his brothers' future. Could time dilatation serve as something like the mythological fountain of youth? That remains a matter of opinion.

As early as 1905, Einstein revealed the time dilatation of watches. In 1911, he extended his observation to living organisms by comparing their lifetimes—in the spirit of the twin paradox. While the organism at rest would have given way to future generations, the same time span would have passed in just an instant for an organism traveling at high speed (which then returns to its starting point). According to Einstein, Louie remains younger for just one reason: because his own time has to slow down relative to a resting observer. In other words, his "biological clock" (proper time) ticks slower than the biological clock of its brothers (Earth time) for the entire duration of his journey in outer space. He only traveled into the future because he did not live at such high speed, so to speak. When his velocity has reached that of light, it follows that Louie's own time would be exactly zero. Time would stand still! Fortunately, this will never happen, because objects with a rest mass (like atoms, and consequently also humans or ducks) can never reach the speed of light exactly. All the universe's energy would be necessary. No, not even that would be enough. The latter, by the way, can again be attributed to a significant degree to the effect of time dilatation. Suppose Louie wanted to accelerate to the speed of light. Then, his inertia (resistance of mass against acceleration) would increase more and more until in the end it would be infinite. The actual reason for this lies in the relativistic increase of inertial mass. Mathematically, it can be derived for example by considering an impulse using time dilatation. No matter how much energy the rocket's propulsion would mobilize, it wouldn't make any difference in the end, because the mass and thus the inertia would increase in the same proportion. It's therefore intuitive to understand that there must be proportionality between energy and mass. This expressed in Einstein's world-famous equation $E = mc^2$.

## The Significance for the Quantum Internet

Now, the reader may rightly ask what all this has to do with a quantum internet. Let's start with an important point. As shown above, Einstein's special theory of relativity (STR) is based significantly on the principle that classical (and therefore usable) information can never be transmitted faster than the speed of light. On the other hand, since STR is not just a theory (as some people still like to believe), but a fact proven millions of times, this very important axiom should be evident even for non-physicists. What is its significance for the quantum internet?

1. The classical internet transmits only usable information, which is why the speed with which classical bits can be transferred is limited to the speed of light at maximum. This truism is known to every communications engineer. However, many people do not know that its origin is STR. The quantum internet, on the other hand, makes it possible to transmit quantum information immediately, without delay. This leads to the reverse conclusion that the transmitted information automatically becomes unusable. And that in turn implies that it has to be strictly separated from classical information.

2. However, precisely because qubits are transmitted instantaneously (see quantum teleportation) and because above the transferred information cannot be usable (see point 1 above), their values that are measured later must inevitably remain unknown at first. Otherwise, they would immediately turn into classical information. Therefore, measurements on entangled qubits are always objectively random automatically. Of course, this quantum randomness cannot have a cause, since it would be possible in principle to know this cause. Therefore, the quantum randomly generated bits in QKD are the very best random numbers that are possible. They can't have a cause, not even in principle. Therefore, they can never be generated algorithmically.

3. Since qubits always remain "phantoms" for reasons of physics, it follows that any eavesdropping operation is

impossible. Classical information however allows such hacker attacks, for which at least a doubling of the information is required. It follows that for nonclassical quantum information, the reverse must be true. That is why qubits maintain a security that is intrinsic. They also cannot be duplicated (no-cloning principle). That is a necessary, however not yet sufficient criterion. This is why its validity has to be demonstrated by an exact quantum mechanical proof (see below).

4. The intrinsic security that follows from points 1–3 above would only be directly at risk if—in absolute contradiction to the STR—superfast transmission of classical information would be possible *after all*. In particular, that would cast considerable doubts on the validity of the no-cloning principle.

## 3.6   The No-Cloning Theorem

The key to safety in the quantum internet is the no-cloning theorem. This applies in particular to the technology, which has to break new ground for the development of quantum repeaters or of error correction methods that are suitable for quantum processes. The unlimited validity of the no-cloning principle is now being questioned by numerous critics, particularly with regard to its capacity to guarantee inherent security based on the fundamental laws of physics. Such judgments can be countered by the fact that the theorem can be concluded logically from fundamental assumptions in theoretical physics. Such evidence was first provided in 1982 by William Wootters et al. This consists in a proof of contradiction. The point of departure is the assumption that a quantum mechanical procedure exists which can copy arbitrary qubits perfectly. This statement is then contradicted using common operator mathematics. The existence of the no-cloning theorem proves to be a consequence of the unity of time evolution operators, which in turn results directly from the axioms of quantum mechanics.

Now, if you choose not to place your trust in the mathematical prowess and the logical abilities of these renowned theorists, perhaps you will be convinced by another, quite different, reason, namely Einstein's theory of relativity. As explained above, the special theory of relativity is based on two basic assumptions. The first is that inertial systems have equal rights. The second is that any signal effect that is faster than light is completely impossible. The latter was taken as the basis for another proof of the no-cloning theorem. Its author was the US physicist Nick Herbert, who theoretically developed a mechanism based on quantum entanglement that is able to transmit useful information faster than light. In his article, Herbert invited his fellow scientists to produce evidence that disproves his thought experiment.

### High Speed Data Transfer?

Of course, it would be a very fine thing for a quantum internet if entanglement could transfer usable information faster than the speed of light. Imagine a quantum satellite creating an entangled channel between Alice and Bob. If Alice then, for example, measures (deliberately, i.e. determined by herself) a binary "1", then Bob at the same moment, an arbitrary distance away, also receives a "1" in his measurement. If she deliberately measures a "0", then immediately Bob measures a "0" as well, and so on. In this way, Alice could transmit digital information to Bob at hyper-speed. Something like that would transcend any known dimension of communications engineering. Unfortunately, though, this is not possible. As has already been shown in detail in the example of QKD, qubits decay into quantum random bit values with every measurement, just as every measurement on entangled systems is subject to objective randomness. This means that information can never be transmitted directly in this way. Also, Einstein's theory of relativity would otherwise collapse like a house of cards. After all, experimentally confirmed time dilatation results as direct consequence of this impossibility.

## Nick Herbert's "Superluminal Device"

Nick Herbert's thought experiment addresses just this question. Wouldn't it be possible to invent a hyper-light fast data transfer machine on the basis of the entanglement? It's important to know that in 1982, laser physics had not yet been thoroughly investigated. The principle of the laser is based on light amplification by stimulated emission. At the time, it was not clear whether it was possible to copy certain quantum states perfectly multiple times. If that were the case, then one could send an input state into a laser, and the same state would emerge at the output in multiple copies. It would then become possible to introduce a single photon with a very specific state into the laser and output trillions of photons with exactly the same properties.

Herbert therefore hypothetically proposed something he called a FLASH system. The acronym stands for "first laser marketed superluminal connection". The basic concept is simple. When Alice and Bob do a measurement in an entangled two-photon system, then (as soon as Alice has her measurement value) the measured value of Bob is also predetermined. Because of quantum randomness, however, no usable information can be transferred (see above). But if the photon, before it reaches Bob, passes through the laser as an input state, then it leaves the laser as trillions of copies. The advantage here is that Bob can then divide this laser beam with the help of a beam splitter and, using "photon statistics", determine unambiguously which state Alice has just prepared. For example, if Alice sets this state to "1", and another unknown state to "0", then she can use a particular sequence of settings to transfer classical information to Bob. As a result of quantum entanglement, information is transferred faster than the speed of light.

All in all, the system works like a telegraph. Alice selects her measurement as if they were the dots and dashes of a Morse code. Bob would be able to decipher every bit of her code at high speed. For this case, the key is to assume that quantum states can be reproduced perfectly.

**More Details**

Imagine a source that emits entangled photons in opposite directions, like in our EPR experiments. The photons can be endowed with either linear or circular polarization. Circular polarization (= spin of the photon) means that the electric field vector strength describes a circular motion along the propagation direction of the light wave. This can be done clockwise or counterclockwise. In order to present the situation as simply as possible, let's introduce a few abbreviations: linearly polarized light (Lp) is H = horizontally polarized or V = vertically polarized and circularly polarized light (Ci) is R = right-circularly polarized or L = left-circularly polarized. Now the correlations are presented as follows. If Alice measures H, then Bob according to the Bell theorem will measure V. If Alice measures R, then Bob measures L. Consequently, Alice's measurement has an immediate effect on Bob's measurement of his photon. The effect is faster than light, as a result of the entanglement as described above. Alice, of course, has free choice of whether to measure linear or circular polarization. Suppose Alice chooses circular polarization and randomly measures L, then, as a result of the entanglement, Bob's state is automatically determined as R. Now Herbert assumes that this photon, before it arrives in Bob's measurement device, passes said laser as an input state. Under the hypothesis that the input state laser equals its output state, a beam of R-polarized photons would then reach Bob. With a beam splitter, he would be able to divide the beams and measure linear polarization on one half of the photons and circular polarization on the other half. Herbert concluded that in this case, 50% of the photons were purely in state R and 25% in H and in V respectively. Based on this measurement result, Bob would know for certain that Alice measured circular polarization. This means that entanglement, which by definition is instantaneous, would be able to transfer the single bit directly at the speed of light. Let's take an example. Alice wants to transmit the bit sequence 1 0 0 1. She therefore chooses her measurement settings Ci, Lp, Lp, Ci…. If Bob knows how the bits are assigned, he is able to reconstruct the sequence without ambiguity.

## Theorists Refute the Thought Experiment

It's too bad that Herbert's superluminal device would not work in practice. As William Wootters, Wojciech Zurek, Tullio Weber, Giancarlo Ghirardi and Dennis Dieks have made plain, such a device would not be able to send signals faster than the speed of light. The reason is that a photon in the state R exists only as a linear combination of the states H and V. Each of these sub-states would be amplified in the laser. The output, therefore, won't be a pure R-state, but a superposition of two states, one where all photons are in state H and one where they all are in state V, with a 50% probability for each of the two alternatives. Bob would therefore only receive "noise". He would never be able to get a state where R would be 25% H and 25% V (since

50% in H or 50% in V). Conclusion: Herbert's FLASH system would not work—and both STR and no-cloning theorem remain valid.

## A Fundamental Discovery

It is interesting to note that it was Herbert's thought experiment which got theorists interested in the subject. And only as a result of that interest was the significant no-cloning principle discovered. Herbert, when developing his device, assumed that quantum information is copied by the laser. However, the quantum mechanical proof later showed that it is not possible to copy any arbitrary quantum state without destroying the original one. The second important conclusion is that quantum mechanics and special relativity are consistent with each other (not only with respect to this). And that's very lucky, because otherwise unspeakable paradoxes would arise. Suppose it were possible to copy quantum information. It would then be possible by implication to transfer information at superluminal velocity. It can be shown that the theory of relativity's principle of causality would then be turned upside down. Since (professionally speaking) the order of spatial events depends on the observer, this can lead to problems of causality. This is because if in one reference system, event A occurs prior to event B, but B occurs before A in the other systems, then it both follows that A can be the cause of B and B can be the cause of A. This leads to paradoxes in which an event retroactively prevents itself in the past. At the same time however, it is possible to develop scenarios that make time travel into the past conceivable at superluminal velocity. The logical consequence is the STR's demand that only events that are "time-like" or "light-like" to each other can influence each other without causality problems arising. For this reason, the STR axiomatically assumes that the speed of light has a maximum value (approximately 300,000 km/s), independent of the movement of the light source and the observer, which can never be exceeded. Since light also has a signaling effect and consequently transfers information, this assumption also applies synonymously to usable information. Einstein was already plagued with such problems, for they would result in a series of curious peculiarities such as the famous "grandfather paradox": If I were able to travel back in time to shoot my grandfather dead, then my mother would not have been born, and I would not even exist. For some reason, nature is always making sure that something so paradoxical never happens in our world. Apparently, part of this are appropriate modifications of quantum theory with laws such as no-cloning or objective randomness.

## 3.7   Closing Remarks

The reader has now become acquainted with an outline of the physics playground where the technology of the quantum internet may emerge one day. With this, the most important part of the requirement which brings the vision to its realization is fulfilled. It's a well-known truism that only such things can be implemented technologically that lie within the realms of nature's laws. In this case, nature shows herself from her most generous side by significantly expanding the perspectives of information theory. The future quantum internet not only exchanges security based on algorithms with security based on the laws of physics, but also enables the networking of quantum computers for decentralized and modular computations. And that's on top of its hyper-fast capabilities of coordination. Methods based on quantum teleportation are used for this purpose. The associated repeater technology makes it possible to transmit quantum information over long distances, which is then stored in a local quantum memory. This opens up a new dimension for data rates which eclipses anything that has been possible to date. What may seem surprising is that this quantum theory which seems so very weird reflects the actual structure of this world. Consequently, our everyday world is nothing more than the result of an omnipresent decoherence. Today's prevalent view among scientists is that quantum physics has to be the most fundamental description of nature. That is, if it's not part of a superordinate theory from which it emerges as a special case, if not as an axiom. This is particularly important because history has repeatedly shown that fundamental discoveries in physics usually bring about fundamental changes for humanity.

For some readers it may seem disappointing that the quantum internet is ultimately not suitable for streaming, blogging or game downloading. Also our everyday e-mail and online business transactions will continue on the basis of the classic internet. But thanks to quantum technology, data protection—especially for the transactions described above—will accomplish unprecedented quality levels. The primary concern here is the long-term security of IT systems, as will be necessary for financial transactions, modern industry concepts, critical infrastructures, central management systems and future mobile communications requirements. Ultimately, society as a whole will benefit from this. Once again, it should be noted that the current concept our security is based on cannot be a permanent solution, simply due to its algorithmic complexity. Improvements in computer performance may occur very suddenly and cause abrupt changes. This requires timely research in the area of algorithms and implementations, including both classical and quantum mechanical methods. QKD has proven to be particularly promising in combination with such methods. Quantum communication can also be used for a number of other procedures such as security seal scheduling or quantum authentication. And research will not stop here.

The second key aspect of the quantum internet is, technologically speaking, a more distant objective: the networking of powerful quantum computers. Already today, quantum simulators show very promising potential. They not only enable profound insights into the world of complex multi-particle systems, they also provide solutions for any quantum problem that cannot be calculated efficiently by a classical computer. With the discovery of topological materials, new opportunities are emerging for solid-state computing. The knowledge gained may lead to innovative technologies that benefit humanity. The cooperation of experimental physicists, theorists and specialists from technology and industry will very likely lead to the realization of a technically usable quantum computer. This is probably only a question of time and available resources. According to today's assessment, the first real breakthrough is expected within the next 10–20 years. However, the development will become really interesting only around 2050, picking up speed rapidly around that time. It is easy to imagine that from then on, many people will use the possibility of an original quantum cloud. Programmers all over the world are supporting the process with successive optimizations. All this may result in truly innovative aspects and applications which nobody has been able to even imagine today. In any case, quantum computers require a differentiated approach and a new way of thinking that has nothing to do with the classical way of programming. A new era is dawning. Already today, "Think quantum!" is the motto of Google's research lab. This might become one of the next generation's leitmotifs. Many young and creative people will participate in this process, driving development. Every child born today could already be directly affected by this new era of computers and the internet.

When CERN physicist Tim Berners-Lee invented the World Wide Web in the late 1980s, he could not have imagined the global revolution that would cause. Physicists will continue to sustainably influence the fate of humanity by providing key inputs for highly innovative technologies. According to Hoimar von Ditfurth, the task of physics is to explain the world without miracles. To achieve this, science had to create quantum theory. And its implications are miraculous enough. This is also true of the technological marvel this book is about, a masterpiece of science that does not have to remain fiction: the universal quantum hypernet.

# References

Einstein, A.: Quantenmechanik und Wirklichkeit. Dialectica **2**, 320–324 (1948)

Einstein, A., Podolsky, B., Rosen, N.: Can quantum-mechanical description of physical reality be considered complete. Phys. Rev. **47**, 777–780 (1935)

Herbert, N.: FLASH—a superluminal communicator based upon a new type of quantum measurement. Found. Phys. **12**, 1171 (1982)

https://www.youtube.com/watch?v=Pf92k-sfKdk&t=1349s

https://www.youtube.com/watch?v=XirbbUxOxiU

https://arxiv.org/abs/1103.3566

https://arxiv.org/abs/1801.06194

https://doi.org/10.1103/PhysRevLett.49.1804

https://arxiv.org/abs/quant-ph/9810080

https://de.wikipedia.org/wiki/Ubiquitous_computing

https://www.oeaw.ac.at/detail/event/pan-jianwei-unter-top-ten-forschern/

https://www.nature.com/articles/nphoton.2017.107

https://de.wikipedia.org/wiki/Schrödingers_Katze

Wootters, W., Zurek, W.: A single quantum cannot be cloned. Nature **299**, 802 (1982)

© Springer Nature Switzerland AG 2020
G. Fürnkranz, *The Quantum Internet*,
https://doi.org/10.1007/978-3-030-42664-4

Zeilinger, A.: Einsteins Schleier, p. 171. C. H. Beck München, Munich (2003)

Zeilinger, A.: Einsteins Spuk, p. 86. C. Bertelsmann München, Munich (2005a)

Zeilinger, A.: Einsteins Spuk, p. 73. C. Bertelsmann München, Munich (2005b)

Zeilinger, A.: Einsteins Spuk, 201ff. C. Bertelsmann München, Munich (2005c)

# Index

© Springer Nature Switzerland AG 2020
G. Fürnkranz, *The Quantum Internet*,
https://doi.org/10.1007/978-3-030-42664-4